Water Science and Technology Library

Volume 74

The aim of the Water Science and Technology Library is to provide a forum for dissemination of the state-of-the-art of topics of current interest in the area of water science and technology. This is accomplished through publication of reference books and monographs, authored or edited. Occasionally also proceedings volumes are accepted for publication in the series.

Water Science and Technology Library encompasses a wide range of topics dealing with science as well as socio-economic aspects of water, environment, and ecology. Both the water quantity and quality issues are relevant and are embraced by Water Science and Technology Library. The emphasis may be on either the scientific content, or techniques of solution, or both. There is increasing emphasis these days on processes and Water Science and Technology Library is committed to promoting this emphasis by publishing books emphasizing scientific discussions of physical, chemical, and/or biological aspects of water resources. Likewise, current or emerging solution techniques receive high priority. Interdisciplinary coverage is encouraged. Case studies contributing to our knowledge of water science and technology are also embraced by the series. Innovative ideas and novel techniques are of particular interest.

Comments or suggestions for future volumes are welcomed.

Vijay P. Singh, Department of Biological and Agricultural Engineering & Zachry Department of Civil Engineering, Texas A and M University, USA Email: vsingh@tamu.edu

More information about this series at http://www.springer.com/series/6689

Baxter E. Vieux

Distributed Hydrologic Modeling Using GIS

Third Edition

 Springer

Baxter E. Vieux
Professor Emeritus
School of Civil Engineering and
 Environmental Science
University of Oklahoma
Norman, OK
USA

and

Principal, Vieux & Associates, Inc.
Norman, OK
USA

Software: Additional materials are available on the software website (via the Help menu) including model description, tutorials and data sets. Vflo® software may be downloaded with an evaluation license from: www.vieuxinc.com/getvflo.

ISSN 0921-092X   ISSN 1872-4663   (electronic)
Water Science and Technology Library
ISBN 978-94-024-0928-4   ISBN 978-94-024-0930-7   (eBook)
DOI 10.1007/978-94-024-0930-7

Library of Congress Control Number: 2016946000

Printed on acid-free paper

This Springer imprint is published by Springer Nature
The registered company is Springer Science+Business Media B.V. Dordrecht

To my wife, Jean and to our children,
William, Ellen, Laura, Anne, and Kimberly,
and to my parents.

Foreword

"Distributed Hydrologic Modeling Using GIS" presents a thorough examination of distributed hydrologic modeling. Application of distributed hydrologic modeling is now an established area of practice. The increased availability of sufficiently detailed spatial data and faster, more powerful computers has motivated the hydrologist to develop models that make full use of such new data sets as radar rainfall and high-resolution digital elevation models (DEMs). The combination of this approach with Geographic Information Systems (GIS) software has allowed for reduced computation times, increased data handling and analysis capability, and improved data display. The twenty-first century hydrologist must be familiar with the distributed parameter approach as the spatial and temporal resolution of digital hydrologic data continues to improve. Additionally, a thorough understanding is required of how this data is handled, analyzed, and displayed at each step of hydrologic model development.

It is in this manner that this book is unique. First, it addresses all of the latest technologies in the area of hydrologic modeling, including Doppler radar, DEMs, GIS, and distributed hydrologic modeling. Second, it is written with the intention of arming the modeler with the knowledge required to apply these new technologies properly. In a clear and concise manner, it combines topics from different scientific disciplines into a unified approach aiming to guide the reader through the requirements, strengths, and pitfalls of distributed modeling. Chapters include excellent discussion of theory, data analysis, and application, along with several cross references for further review and useful conclusions.

This book tackles some of the most pressing concerns of distributed hydrologic modeling: What are the hydrologic consequences of different interpolation methods? How does one choose the data resolution necessary to capture the spatial variability of your study area while maintaining feasibility and minimizing computation time? What is the effect of DEM grid resampling on the hydrologic response of the model? When is a parameter variation significant? What are the key aspects of the distributed model calibration process?

In "Distributed Hydrologic Modeling Using GIS," Dr. Vieux has distilled years of academic and professional experience in radar rainfall applications, GIS, numerical methods, and hydrologic modeling into one single, comprehensive text. The reader will not only gain an appreciation for the changes brought about by recent technological advances in the hydrologic modeling arena, but will also fully understand how to successfully apply these changes toward better hydrologic model generation. "Distributed Hydrologic Modeling Using GIS" not only sets guiding principles to distributed hydrologic modeling, but also asks the hydrologist to respond to new developments calling for additional research. These new methods have revolutionized the fields of hydrology and floodplain analysis in the past 15 years and created new and amazing data and models that will have to be taught to a whole generation of scientists and engineers. All of the above make this a unique, invaluable book for the student, professor, or hydrologist seeking to acquire a thorough understanding of this area of hydrology.

Philip B. Bedient
Herman Brown Professor of Engineering
Department of Civil and Environmental Engineering
Rice University, Houston, TX, USA

Preface

I wanted to write this book on distributed hydrologic modeling, from a spatial perspective. When the modeling approach seeks to preserve "distributed" characteristics, then geospatial information management becomes important, particularly in the setup and assignment of parameters, and in related actions involving query, manipulation, and analysis using a geographic information system (GIS). All models are an abstraction from actual hydrologic processes. Longstanding representation by lumping of parameters at the watershed or river basin scale was originally necessitated by a lack of information, and limited computer and data resources. With available geospatial data sets for soils, topography, land use, and precipitation, there is a need to advance the science and practice of hydrology, by capitalizing on these rich sources of information.

To advance from lumped to distributed representations requires re-examination of how we model for both engineering purposes and scientific understanding. We could reasonably ask what laws govern the complexities of all the paths that water travels, from precipitation falling over a river basin to the flow in the river. We have no reason to believe that each unit of water mass is not guided by Newtonian mechanics, making conservation laws of momentum, mass, and energy applicable. Once we embark on fully distributed representations of hydrologic processes, we have no other choice than to use conservation laws (termed "physics-based") as governing equations. It is my conviction that hydrologists will opt for distributed physics-based representation of hydrology, because it has a firmer scientific foundation than traditional lumped conceptual techniques, and takes advantage of a wealth of geospatial data available within a GIS framework.

What was inconceivable a decade ago is now commonplace in terms of computational power; availability of high-resolution geospatial data; and management systems supporting detailed mathematical modeling of complex hydrologic processes. Technology has enabled the transformation of hydrologic modeling from lumped to distributed representations with the advent of new sensor systems such as radar and satellite, high-performance computing, and orders-of-magnitude increases in storage. Global remote sensing data sets now are available at 30 cm resolution, and soil moisture estimates from satellite at 500 m. Such tantalizing geospatial detail could be of use in making better hydrologic predictions or estimates of the extremes of weather, drought, and flooding, but only if we adapt new modeling techniques that can leverage such detail.

When confronted with the daunting task of modeling a natural process, individuals may be ill-equipped to address even a few of the most important aspects affecting hydrologic processes. In actuality, water does not care whether it is flowing through a meteorologist's domain or that of a soil scientist's. Early in my training, I realized that the flow direction grid derived from digital terrain, could be used to create a system of equations solving channel and overland flow. Or that a soil map could be reclassified to produce runoff curve numbers for calculating rainfall-runoff from a watershed useful in the design of flood control dams or reducing erosion and sedimentation. Applying these "new" distributed hydrologic methods and techniques derived from diverse scientific domains seemed natural, if only because the common fabric linking them together was the physics of natural processes that govern the distribution of water (or lack thereof) on or near the earth's surface.

The writing of this book attempts to balance between principles of distributed hydrologic process modeling on the one hand, and how modeling can be implemented using GIS. As the subject emerged during the writing of this book, it became clear that there were issues with geospatial data formats, spatial interpolation, and resolution effects on information content or drainage network detail that could not be omitted. Examples and case studies are included that illustrate how to most effectively represent the process, while avoiding the many pitfalls inherent in such an undertaking. It is my hope that this monograph provides useful guidance and insights to those hydrologists interested in physics-based distributed hydrologic modeling.

This third edition has updated reference citations; additional figures and tables; and needed corrections. Case studies are provided that demonstrate principles of distributed physics-based hydrology. Many of the examples and case studies provided rely on distributed hydrologic model software, V*flo*®, for which I guided development.

Norman, OK, USA   Baxter E. Vieux

Acknowledgments

I wish to thank my colleagues who contributed greatly to the writing of this and prior editions of this book. Over the course of many years, I have enjoyed collaborations with colleagues that have encouraged the development and application of distributed modeling. Special thanks go to the team at Vieux & Associates, Inc., especially Ryan Hoes, Jennifer French, Edward Koehler, Brian McKee, and Jean Vieux for support and assistance with hydrometeorology and V*flo*® software development.

Contents

Chapter 1
Introduction to Physics-Based Distributed Hydrology

Abstract The spatial and temporal distribution of the inputs and parameters controlling surface runoff can be managed efficiently within a GIS framework. Examples include maps describing slope and drainage direction, land use/cover, soil parameters such as porosity or hydraulic conductivity, rainfall, and meteorological variables controlling evapotranspiration. The subject of this book is how these maps of geospatial information can be harnessed to become model parameters or inputs defining the hydrologic processes of surface and subsurface runoff. As soon as we embark on the simulation of hydrologic processes using GIS, the issues that are the subject of this book must be addressed.

1.1 Introduction

Distributed hydrologic modeling has become an accepted approach for a variety of applications. Simultaneous advances in computing power and hydraulic/hydrologic modeling technology make it possible to leverage high-resolution data sources now available. New instrumentations such as Laser Imaging and Ranging (LIDAR), land use/cover interpreted from satellite remote sensing and RADAR measurement of precipitation provide more detail than ever before. When geospatial data are used in hydrologic modeling, important issues arise such as the necessary resolution to capture essential variability, or derivation and regionalization of model parameters that are representative of the watershed. It is not surprising that Geographic Information Systems (GIS) have become an integral part of hydrologic studies considering the spatial character of parameters and precipitation controlling hydrologic processes. The primary motivation for this book is to bring together the key ingredients necessary to effectively model hydrologic processes in a distributed manner. Often there are only sparse streamflow observations making it all the more necessary to incorporate the *physics* of the processes, rather than develop ad hoc regression relationships between precipitation and runoff that lack transferability.

Historical practice has been to use lumped representations because of computational limitations or because sufficient data were not available to populate

© Springer Science+Business Media Dordrecht 2016
B.E. Vieux, *Distributed Hydrologic Modeling Using GIS*,
Water Science and Technology Library 74, DOI 10.1007/978-94-024-0930-7_1

a distributed model database. The number of discrete elements used to represent processes determines the degree to which we classify a model as lumped or distributed. Several distinctions of terms are used in this book and by researchers or model developers such as subbasin lumped models are intrinsically lumped, though at the subbasin level. Such models rely on conceptual "buckets" that fill or drain based on simplified relationships or on regression equations such as unit hydrograph methods. But these are not *physics-based* because the basis is not the conservation equations of mass energy or momentum. Some models are described as being *physically-based* because notions of upper root zone moisture and lower root zone moisture may have some basis in reality. However, such parameterizations are not based on physics but rather on conceptual simplifications that can only be identified through calibration with an observed streamflow containing sufficient information content. A further tell-tale sign of a conceptual model is that the parameters have no basis in physical reality and must be changed from season to season, making calibration valid only for the specific period of calibration. Some equations are only valid for a finite volume but not at a point. For example, Darcy's law that governs subsurface flow is only conceivable for a volume for which porosity can be defined, which is incompatible at smaller scales, say at the pore space scale. In a distributed physics-based approach, the discrete element used is called an averaging volume and admittedly is lumped at the sub-grid scale.

Whether representation of hydrologically homogeneous areas can be justified depends on the uniformity of the hydrologic parameters representing the terrain. The watershed in Fig. 1.1 is a 25.6 km^2 area that drains from steeply sloping foothills of the Colorado Rocky Mountains, onto a flatter through highly urbanized area. Such drainage areas along the Front Range can produce damaging floods from intense precipitation typical of the region. The upper portion consists of steeply sloping mountainous terrain with natural vegetative cover, grass, and forest (green). The lower portion becomes increasingly urbanized with impervious surfaces associated with commercial, residential, and transportation land uses but deeper more pervious soils are not covered by impervious surfaces (light gray). Lumping this watershed into one, or only a few subbasins could distort the hydrologic behavior and not represent runoff well, especially when highly variable precipitation falls in the steeper undeveloped portion, or vice versa, when the highest intensities fall over urban portions of the watershed.

Some of the effects associated with lumped approaches to hydrology include

1. The resulting model is not physics-based and depends on empirically derived timing parameters to generate or route runoff through a lumped subbasin network.
2. When historic streamflow is necessary for deriving model parameters, then lumping is necessary at locations where observations exist.
3. When the number of subbasins or their arrangement is changed, unexplained changes in hydrologic response require recalibration.

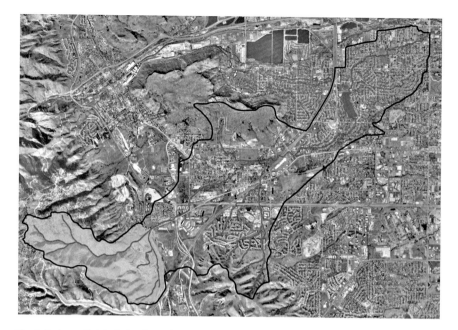

Fig. 1.1 Lena Gulch drains from the Colorado foothills in the west (*green*) to urbanized areas

4. Parameter variability may not be represented accurately by lumping at the subbasin scale.
5. A gridded drainage network supports hydrologic prediction in any grid cell without having to re-delineate the watershed to new locations where modeled hydrographs are desired.

1.2   Model Classification

It is useful to consider how physics-based distributed (PBD) models fit within the larger context of hydrologic modeling. Figure 1.2 shows a schematic that helps in classifying modeling approaches.

Deterministic is distinguished from *stochastic* in that a deterministic river basin model estimates the response to an input using either a conceptual mathematical representation or a physics-based equation. Conceptual representations usually rely on some type of linear reservoir theory to delay and attenuate the routing of runoff generated. Runoff generation and routing are not closely linked and therefore do not interact. Physics-based distributed (PBD) models use equations of conservation of mass, momentum, and energy to represent both runoff generation and routing in a linked manner. Following the left-hand branch in the tree, the distinction between runoff generation and runoff routing is somewhat artificial, because they are

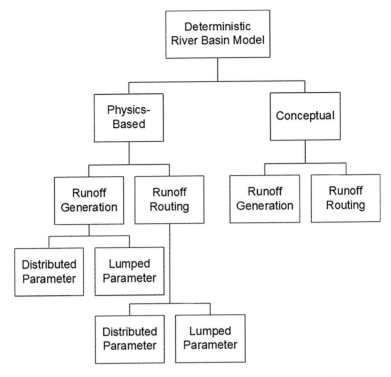

Fig. 1.2 Model classification according to approach and distributed/lumped distinctions

intimately linked in most distributed model implementations. However, by making a distinction we can introduce the idea of lumped versus distributed parameterization for both overland flow and channel flow. A further distinction is whether overland flow or subsurface flow is modeled with lumped or distributed parameters.

The degree of interconnection of subsurface runoff and the surface may be expressed in different ways. Either when the subsurface travels horizontally through an aquifer or if saturation excess runoff is routed as surface drainage to stream channels. In V$flo^{®}$, when the soil profile saturates, the runoff (100 % of rainfall) then runs off the cell horizontally according to Manning hydraulics. Alternatively, the subsurface flow moving horizontally through a shallow aquifer can be modeled using the Boussinesq equation solved with linearization (Verhoest and Troch 2000). Combining surface and subsurface flow in a single kinematic wave approximation and routing over a TIN surface was described by Tachikawa et al. (2007); and Vivoni et al. (2004). Various degrees of lumping can be implemented when routing flow through the channels that distinguishes whether uniform or spatially variable parameters are applied in a given stream segment. For example, if a constant routing parameter is used between stream gauges, such as celerity in the Muskingum-Cunge equation, then the timing of the flood wave does not change within the reach and would be considered lumped at that level. Whereas in a fully distributed routing scheme, the velocity could be allowed to vary in each grid cell and thus would be considered a distributed routing scheme.

Determining an appropriate resolution for capturing the essential information contained in a parameter map is investigated in Chap. 4. The spatial resolution used to represent spatially variable parameters is another form of lumping. Changing spatial resolution of data sets requires some scheme to aggregate parameter values at one resolution to another. Resampling is essentially a lumping process, which in the limit, results in a single value for the spatial domain. Resampling a parameter map involves taking the value at the center of the larger cell and then averaging or by another operation. If the center of the larger cell happens to fall on a low/high value, then a large cell area will have a low/high value.

Resampling rainfall maps can at first produce noticeable sampling effects at a small resolution, yet produce erratic results as the resolution increases in size. Farajalla and Vieux (1995) and Vieux and Farajalla (1994) applied information entropy to infiltration parameters and hydraulic roughness to discover the limiting resolution beyond which little more was added in terms of information. Over-sampling a parameter or input map at a finer resolution may not add any more information, either because the map, or the physical feature, does not contain additional information. Of course, variations exist physically; however, these variations may not have an impact at the scale of the modeled domain.

Model input with coarse steps can also be considered lumping and can influence the PBD models significantly depending on the size of a basin. This sensitivity is due to the conservation equations being solved that rely on rainfall intensities rather than accumulation. Because unit hydrograph approaches are based on rainfall accumulation, they are less sensitive to changes in the intensity caused using longer forcing time steps. Temporal lumping occurs with aggregation over time of such phenomena as stream flow or rainfall accumulations at 5-min, hourly, daily, 10-day, monthly, or annual time series. Small watersheds may be more sensitive than larger watersheds and could require rainfall time series at 5-min intervals.

Numerical solution of the governing equations in a physics-based model employs discrete elements. The three representative types are finite difference, finite element, and stream tubes. At the level of a computational element, a parameter is regarded as being representative of an average process. Thus, some average property is only valid over the computational element used to represent the runoff process. For example, porosity is a property of the soil medium but it has little meaning at the level of the pore space itself. Thus, resolution also depends on how well a single value represents a grid cell. Sub-grid parameterization should be a consideration for larger grid cells where important variability is "averaged-out".

From a model perspective, a parameter should be representative of the surface or medium at the scale of the computational element used to solve the governing mathematical equations. This precept is often exaggerated as the modeler selects coarser grid cells, losing physical significance. In other words, the runoff depth in a grid cell of 1-km resolution can only be taken as a generalization of the actual runoff process and may or may not produce *physically realistic* model results.

When modeling large basins at fine resolution, computational resources can easily be exceeded, even with modern computing resources. This limitation motivates the need for coarser model resolution than is represented by digital terrain data. The DEM available from USGS used to create the Lena Gulch watershed is at 3-m resolution. A given resolution may be too coarse to represent highly variable parameters such as infiltration or roughness. Besides parameter values not being adequately represented, coarser resolution models will produce more attenuated results mainly due to the slope being reduced but also because fewer finite elements tend to produce lower peaks before equilibrium is approached. Figure 1.3 shows the Lena Gulch watershed (as in Fig. 1.1) but with a model grid at 1,667 × 1,667-m resolution (1 × 1 mile) delineated from a 3-m DEM. Figure 1.4 shows the same watershed but delineated at 100-m resolution. Coarse resolution tends to produce more attenuated hydrographs as illustrated by the simulation with the same parameters, i.e., 100 % impervious and overland cell roughness set to $n = 0.035$ with input from the same storm hyetograph with intensities exceeding 500 mm/h and a depth of 100 mm lasting 2 h. Figure 1.5 shows the two hydrographs produced by the 100-m (blue line) and 1,667-m (red dots) resolution basins. The coarser resolution model peaks at 175 m^3s^{-1}, whereas, the finer resolution model (100 m) peaks at 220 m^3s^{-1}. Note that parameters with the same constant value were used in both, so that the only difference is resolution.

If spatial variation can be sufficiently represented at 100-m resolution or larger, then computational advantages will result because the larger cell size occupies less computer memory and the computational time step will likely be longer. The

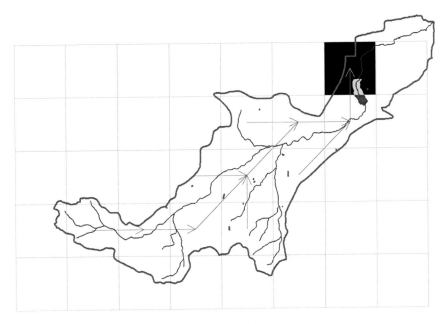

Fig. 1.3 Modeled area shown at 1,667-m (1 × 1 mile) resolution

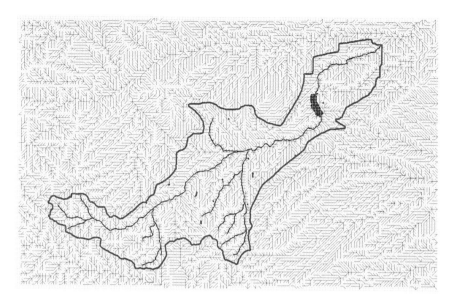

Fig. 1.4 As in Fig. 1.3 above but considerably finer at 100-m resolution

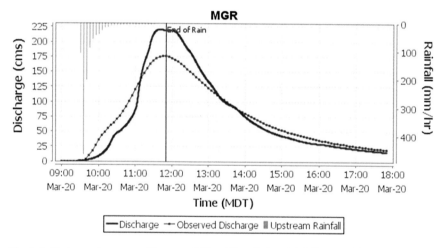

Fig. 1.5 Hydrograph compared between 100-m (*blue line*) and 1,667-m (*red dots*) resolution

infiltration, hydraulic roughness, and terrain slope parameters are particularly sensitive to model resolution. Figure 1.6 shows the slope parameter histogram, which when lumped, averages 15.94 %. The effect of lumping is reduced information content, which is rather high given the broad range of slope values in the histogram. As demonstrated in Fig. 1.7, when the spatially variability is retained, the calibrated model closely reproduces the observed discharge for a peak during

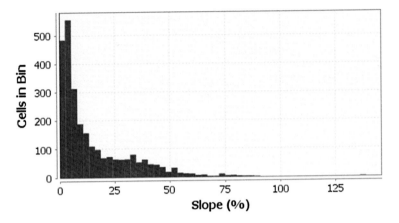

Fig. 1.6 Histogram of spatially distributed slope in the 100-m resolution model

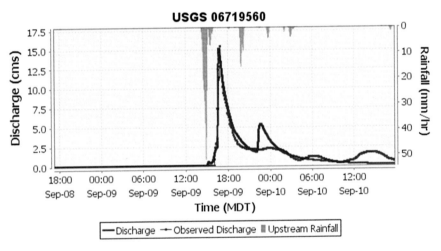

Fig. 1.7 Watershed response modeled with spatially variable parameters at 100 m

the Front Range Flood of 2013 (NWS 2013). When slope, hydraulic roughness, and saturated hydraulic conductivity are lumped, the model does not reproduce the discharge hydrograph adequately as seen in Fig. 1.8.

An important decision faced when setting up a distributed hydrologic model is determining the resolution that captures the parameter variation, while preserving computational efficiency. The watershed response dependency on resolution illustrates why calibration is necessary for a physics-based distributed model and why re-calibration is often required when changing to larger or smaller resolution. Depending on the areal extent of the watershed and the variability inherent in each

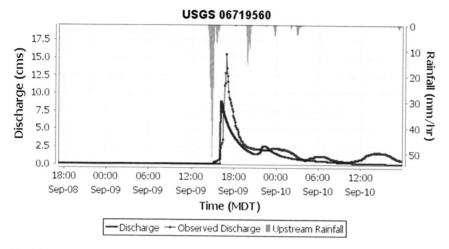

Fig. 1.8 Watershed response modeled with lumped parameters at 100 m

parameter, small variations may not be important while other variations may exercise a strong influence on model performance. Chapter 4 presents methods for calculating information content and its effects on model simulations of watershed response.

1.3   Geospatial Data for Hydrology

In Chap. 2, the major data types necessary for distributed hydrologic modeling are examined. Depending on the particular watershed characteristics, many types of data may require processing before they can be used in a hydrologic model. GIS software is used to assemble and analyze geographic data, available as global data sets, though frequently at coarse resolution. Some geospatial data processing may be necessary before beginning model setup. Hydrologic models are now available that are designed to use geospatial data effectively. Once a particular spatial data source is considered for use in a hydrologic model then we must consider the data structure, file format, quantization (precision), and error propagation. While GIS can be used to process geospatial data, it is often a tedious process since the analysis functions are of general purpose supporting a wider array of applications than hydrology. The relevance of remotely sensed data to hydrologic modeling may not be known without special studies to test whether a new data source provides advantages that merit its use.

1.4   Surface Generation

Several surface generation techniques useful in extending point data to surfaces are described in Chap. 3. Digital representation of terrain requires that a surface be modeled as a set of elevations or other terrain attributes at point locations. Much work has been done in the area of spatial statistics and the development of kriging techniques to generate surfaces from point data. In fact, several methods for generating a two-dimensional surface from point data may be enumerated

- Linear interpolation
- Local regression
- Distance weighting
- Moving average
- Splines
- Kriging

The problem with all of these methods when applied to smoothing varying fields such as rainfall, groundwater flow, wind, temperature, or soil properties is that the interpolation algorithm may violate some physical aspect. Gradients may be introduced that are a function of the sparseness of the data and/or the interpolation algorithm. Values may be interpolated across distinct zones where natural discontinuities exist.

Suppose, for example, that several piezometric levels are measured over an area and that we wish to generate a surface representative of the piezometric levels or elevations within the aquifer. The inverse distance weighting (IDW) scheme is commonly used but almost certainly introduces artifacts of interpolation that violate physical characteristics, viz., gradients are introduced that would indicate a flow in directions contrary to the known gradients or flow directions in the aquifer. In fact, a literal interpretation of the interpolated surface may indicate that, at each measured point, pressure decreases in a radial direction away from the well location, which is clearly not the case. Similarly, when IDW is applied to point rain gauge depths, it will appear that rainfall falls mainly at the gauge and diminishes with distance from the gauge.

None of the above methods of surface interpolation are entirely satisfactory when it comes to ensuring physical correctness in the interpolated surface. Depending on the sampling interval, spatial variability, physical characteristics of the measure, and the interpolation method, the contrariness of the surface to physical or constitutive laws may not be apparent until model results reveal intrinsic errors introduced by the surface generation algorithm. Chapter 3 deals with surface interpolation and hydrologic consequences of interpolation methods.

1.5 Spatial Resolution and Information Content

Chapter 4 provides an overview of information theory with an application showing how information entropy is descriptive of spatial variability and its use as a statistical measure of resolution impacts hydrologic parameters such as slope. How resolution in space affects hydrologic modeling is of primary importance. The resolution that is necessary to capture the spatial variability is often not addressed in favor of simply using the finest resolution possible. It makes little sense, however, to waste computer resources when a coarser resolution would suffice. We wish to know the resolution that adequately samples the spatial variation in terms of the effects on the hydrologic model and at the scale of interest. This resolution may be coarser than that dictated by visual aesthetics of the surface at fine resolution.

The question of which resolution suffices for hydrologic purposes is answered in part by testing the quantity of information contained in a data set as a function of resolution. We can stop resampling at coarser resolution once the information content begins to decrease or be lost. Information entropy, originally developed by communication engineers, can test which resolution is adequate in capturing the spatial variability of the data (Vieux 1993).

1.6 Infiltration

Infiltration modeling that relies on soil properties to derive the Green and Ampt equations is considered in Chap. 5. The two basic flow types are: *overland flow*, conceptualized as a thin sheet flow before the runoff concentrates in recognized channels and *channel flow*, conceptualized as occurring in recognized channels with hydraulic characteristics governing flow depth and velocity. Overland flow is the result of rainfall rates exceeding the infiltration rate of the soil. Depending on soil type, topography, and climatic factors, surface runoff may be generated either as infiltration excess, saturation excess, or in combination throughout a watershed. Loague et al. (2010) argued that rainfall runoff modeling can be better achieved by not being overly prescriptive as to assumed mechanisms of Horton (infiltration rate excess) or Dunne-type (saturation excess) runoff and asserted that there is a third type, called 'Dunton', that contains elements of both. Therefore, estimating infiltration parameters from soil maps and associated databases is important for quantifying infiltration at the watershed scale.

The infiltration rate excess first identified by Horton is typical in areas where the soils have low infiltration rates and/or the soil is bare. Raindrops striking bare soil surfaces break up soil aggregates, allowing fine particles to clog surface pores. A soil crust of low infiltration rate occurs particularly where vegetative cover has been removed exposing the soil surface. Infiltration excess is generally conceptualized as a flow over the surface in thin sheets. Model representation of overland flow uses this concept of uniform depth over a computational element though it differs from reality, where small rivulets and drainage swales convey runoff to the major stream channels.

Richards' equation fully describes this process using principles of conservation of mass and momentum. The Green and Ampt equation (Green and Ampt 1911) is a simplification of Richards' equation that assumes piston flow (no diffusion). Loague (1988) found that the spatial arrangement of soil hydraulic properties at hillslope scales (<100 m) was more important than rainfall variations. Order-of-magnitude variation in hydraulic conductivity at length scales on the order of 10 m controlled the runoff response. This would seem to imply that it is impossible to know infiltration rates at the river-basin scale unless very detailed spatial patterns of soil properties are measured. The other possible conclusion is that not all of this variability is important over large areas. Considering that detailed infiltration measurement and soil sampling are not economically feasible over a large spatial extent, deriving infiltration rates from soil maps is an attractive alternative. Modeling infiltration excess at the watershed scale is more feasible if infiltration parameters can be estimated from mapped soil properties.

1.7 Hydraulic Roughness

Chapter 6 presents an overview of developing the hydraulic parameters necessary for modeling surface runoff. Accounting for overland and channel flow hydraulics over the watershed helps our ability to simulate hydrographs at the outlet. In rural and urban areas, hydraulics governs the flow over artificial and natural surfaces. Frictional drag over the soil surface, standing vegetative material, crop residue, rocks lying on the surface, raindrop impact, and other factors influence the hydraulic resistance experienced by runoff. Hydraulic roughness coefficients caused by each of these factors contribute to total hydraulic resistance.

A detailed measurement of hydraulic roughness over any large spatial extent is generally impractical. Thus, reclassifying a GIS map of land use/cover into a map of hydraulic roughness parameters is attractive in spite of the errors present in such an operation. Considering that hydraulic roughness is a property that is characteristic of land use/cover classification, hydraulic roughness maps can be derived from a variety of sources. Aerial photography, land use/cover maps, and remote sensing of vegetative cover become a source of spatially distributed hydraulic roughness. Each of these sources lets us establish hydraulic roughness over broad areas such as river basins or urban areas with both natural and artificial surfaces. The goal of reclassification of a land use/cover map is to represent the location of hydraulically rough versus smooth land use types for watershed simulation. Chapter 6 deals with the issue of how land use/cover maps are reclassified into hydraulic roughness and then used to control how fast runoff moves through the watershed.

1.8 Watersheds and Drainage Networks

Drainage networks may be derived from digital elevation models (DEMs) by connecting each cell to its neighbor in the direction of principal slope. DEM resolution has a direct influence on the total drainage length and slope. Too coarse resolution causes an under-sampling of the hillslopes and valleys where hilltops are cut off and valleys filled. Two principal effects of increasing the resolution coarseness are

1. Drainage length is shortened
2. Slope is flattened

The effect of drainage length shortening and slope flattening on the hydrograph response may be compensating. That is, shorter drainage length accelerates arrival times at the outlet, while flatter slopes delay arrival times. The influence of DEM grid-cell resolution is discussed in Chap. 7.

1.9 Distributed Precipitation Estimation

Chapter 8 discusses the measurement and associated uncertainty of spatially distributed precipitation. One of the most important sources of spatially distributed rainfall data is the radar and without it, distributed modeling cannot be accomplished with full efficiency. Spatial and temporal distribution of rainfall is the driving force for both infiltration and saturation excess. Rain gauge networks can be used alone or in conjunction with weather radar. Gauge-adjusted radar rainfall (GARR) is produced using rain gauge accumulations to bias the correct radar. Another term used is Multisensor Precipitation Estimates (MPE). The US National Weather Service (NWS) produces MPE with operational gauge adjustments for bias correction. Network density has important consequences on representativeness of the spatially distributed rain gauge measurement and its accuracy. The WSR-88D (known as NEXRAD) weather radar network mainly covers the continental US with varying density. While its purpose is surveillance and detection of severe weather, it can have applications to hydrology with added quality control. Understanding how a radar system may be used to produce accurate rainfall estimates provides a foundation for application to hydrologic models. In Fig. 1.9 radar-derived precipitation with bias correction is shown for an extreme event that occurred during the Colorado Front Range Flood Event of 2013. The maximum accumulation shown here is over 210 mm (8.3 in.) and caused four flood peaks during the event as repeated storm cells trained over the area. Achieving accurate rainfall at high resolution in space and time is paramount to accurate flood prediction. Resolution in space and time, errors, quantizing (precision), and availability in real-time or post-analysis is taken up in this chapter.

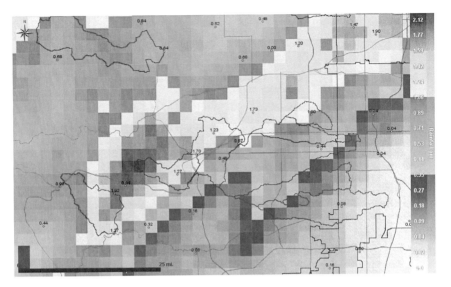

Fig. 1.9 Radar-derived storm total precipitation with bias correction during the Colorado Front Range Flood, September 8–16, 2013

1.10   Surface Runoff Model Formulation

Chapter 9 presents a detailed description of a finite element solution to kinematic wave equations. Physics-based distributed (PBD) models solve governing equations derived from conservation of mass, momentum, and energy. Unlike empirically based models, differential equations are used to describe the flow of water over the land surface or through porous media, or energy balance in the exchange of water vapor through evapotranspiration. In most physics-based models, simplifications are made to the governing equations because certain gradients may not be important or accompanying parameters, boundary, and initial conditions are not known. Linearization of the differential equations is also attractive because nonlinear equations may be difficult to solve. The resulting mathematical analogies are simplifications of the complete form. The full dynamic equations describing the flow of water over the land surface or in a channel may contain gradients that are negligible under certain conditions. In a mathematical analogy, we discard the terms in the equations that are orders of magnitudes less than others. Simplifications of the full dynamic governing equations give rise to zero inertial, kinematic, and diffusive wave analogies. Using simplified or full dynamic mathematical analogies to generate flow rates is a hydraulic approach to hydrology. Using such conservation laws provides the basis in physics for fully distributed models.

If the physical character of the hydrologic process is not supported by a particular analogy, then errors result in the physical representation. Difficulties also arise from the simplifications because the terms discarded may have afforded a complete solution while their absence causes mathematical discontinuities. This is particularly true in the kinematic wave analogy, in which changes in parameter values can cause discontinuities, sometimes referred to as *shock,* in the equation solution. Special treatment is required to achieve a stable solution to the kinematic wave equation, where runoff flows over a surface with spatially variable roughness or slope. Vieux (1988), Vieux et al. (1990) and Vieux (1991) found such a solution using nodal values of parameters in a finite element solution. This method effectively treats changes in parameter values by interpolating their values across finite elements. The advantage of this approach is that the kinematic wave analogy can be applied to a spatially variable surface without numerical difficulty introduced by the shocks that would otherwise propagate through the system. Distributed watershed models that rely on a nodal representation in the numerical solution include *r.water. fea* (Vieux and Gauer 1994) and V*flo*® (Vieux and Vieux 2002).

The full momentum St. Venant equation or at least the diffusive wave analogy is necessary where backwater effects are important. Backwater effects may become important in low-gradient river systems or where detailed bridge hydraulics affects the upstream water surface elevation. The CASC2D (Julien and Saghafian 1991; Julien et al. 1995) is a raster model that uses the diffusive wave analogy to simulate flow on a grid-cell representation of a watershed. The GSSHA model extends the applicability of the CASC2D model to handle surface-subsurface interactions associated with saturation excess runoff on a gridded basis using the diffusive wave analogy. Recent implementation of the WRF-Hydro framework comprised multiple complex process sub-models that routed flow using a one-dimensional diffusive wave analogy (Gochis et al. 2014; Clark et al. 2015a, b). These models solve the diffusive wave analogy using a finite difference grid corresponding to the grid-cell representation of the watershed. The diffusive wave analogy requires additional boundary conditions to obtain a numerical solution in the form of supplying a gradient term at boundaries or other locations. The choice of mathematical analogies depends largely on the gradients of the flow in the river or stream and the type of hydraulic problem that is being solved (Moussa and Bocquillon 1996; Ponce et al. 1978).

The numerical solution presented results in attenuation of the hydrograph. This edition is focused on the distributed hydrologic model called V*flo*®, which uses finite elements in space and finite difference in time to solve the kinematic wave analogy for a watershed composed of gridded parameters and input. The parameters are setup in the model using spatially distributed maps prepared using GIS, which is discussed next.

Physics-based distributed hydrologic modeling relies on conservation equations to create a representation of surface runoff. The kinematic wave mathematical analogy may be solved using a network of finite elements connecting grid cells together. Flow direction in each grid cell is used to layout the finite elements. Solving the resulting system of equations defined by the connectivity of the finite elements provides the possibility of hydrograph simulation at any location in the drainage network. The linkage between GIS and the finite element and finite difference algorithms to solve the kinematic wave equations is examined in detail in Chap. 9. Assembly of finite elements representing the drainage network produces a system of equations solved in time. The resulting solution is the hydrograph at selected stream nodes, cumulative infiltration, and runoff depth in each grid cell.

1.11   Distributed Model Calibration

Chapter 10 addresses these calibration methods where initial estimates of parameter values and their spatial distribution form the framework for model calibration. Once the assembly of input and parameter maps for a distributed hydrologic model is completed, the model must usually be calibrated or adjusted. The argument that physics-based models do not need calibration presupposes perfect knowledge of the parameter values distributed throughout the watershed and of the spatially/temporally variable rainfall input. This is clearly not the case. Besides the parameter and input uncertainty, there are resolution dependencies affecting both parameters and model response. Hydrologists have argued that there are too many degrees of freedom in distributed modeling vis-à-vis the number of observations. This concern does not take into account that if we know the spatial pattern of a parameter, we can adjust its magnitude while preserving the spatial variation. This calibration procedure can be performed by applying scalar multipliers or additive constants to parameter maps until the desired match between simulated and observed is obtained. Advanced methods exploiting the parameter sensitivity of the governing equations can be used to identify optimal parameters through data assimilation (Vieux et al. 1998; White et al. 2002, 2003; Le Dimet et al. 1996, 2009; Sene 2012). The ordered physics-based parameter adjustment (OPPA) method described in Chap. 10 has been particularly adapted to physics-based modeling because it exploits the properties of the governing equations. Predictable parameter interaction and model behavior are hallmarks of the OPPA approach used to identify *physically realistic* distributed parameter values through comparison with streamflow observations.

1.12   Case Studies in Distributed Hydrology

The case studies presented in Chap. 11 illustrate aspects of distributed hydrologic modeling for different watershed scales and timeframe. In Case Study I, the application of physics-based distributed modeling of reservoir inflow forecasting is presented. Reservoir inflow is a key component necessary for efficient power generation and flood management at upstream locations along the impoundment and in downstream river reaches. A hydrometeorological network is operated to measure current precipitation and numerical weather prediction for hydrologic predictions and the potential evapotranspiration (PET) unusual property of being nearly impervious. Its response to rainfall can occur quickly within a few hours. Operational flood forecasts are produced using gauge-adjusted radar rainfall in real-time and for a recent event, flood forecasts are presented. In Case Study III, projections are made relative to GCM projections of climate change impacts on streamflow. A V*flo*® model is calibrated to a historic period with 20 years of multisensor precipitation and streamflow. The model is then used to estimate the watershed response to future precipitation and PET. Streamflow metrics derived from model simulations are used to predict the number and diversity of fish sampling at interior locations throughout the watershed. Streamflow projections at sampling sites are developed by perturbing precipitation and the PET derived from GCM projections, *as-if* the climate has already changed.

1.13 V*flo*®—Software for Distributed Hydrology

Chapter 12 describes model software features and their use in distributed hydrologic prediction and analysis. The formulation of this model supports prediction at scales from small urban catchments up to larger river basins and for areas that have mixed urban and rural land use/cover. The network-based hydraulic approach to hydrology makes it possible to represent both local and main-stem flows with the same model setup and simultaneously. From within the software interface functionality exists to read-in gridded terrain elevation and delineate watershed areas. Because of its importance to physics-based hydrology, import of distributed rainfall input from radar and rain gauge measurements allows the user to benefit from high-resolution input. Event and continuous simulation including soil moisture is supported in both offline and in operational applications. Post-analysis of storm events allows calibration and hydrologic analysis using archived radar rainfall. Advances in modeling techniques; multisensor precipitation estimation; and GIS and remotely sensed data have resulted in an enhanced ability to make hydrologic predictions at distributed locations. The modeling approach described in this book represents a paradigm shift from traditional hydrologic modeling. The V*flo*® software is available for trial use and may be obtained via the Internet, as described at the end of Chap. 12. The V*flo*® software provides Help and Tutorial files that are useful for learning how to setup a distributed hydrologic model and understanding how the model simulates watershed hydrology within a distributed physics-based context.

1.14 Summary

While this book answers questions related to distributed modeling, it also raises others on how best to model distributed hydrologic processes using GIS. Depending on the reader's interest, the techniques described have a wider application than the subset of hydrologic processes addressed in the chapters that follow.

The objective of this book is to present scientific principles of distributed hydrologic modeling. In an effort to make the book more general, techniques described may be applied using many different GIS packages. The material contained in this edition has benefited from more experience with the application of distributed modeling in operational settings and from advances in software development. Advances in research have led to better understanding of calibration procedures and the sensitivity of PBD models to inputs and parameters. The fact that a distributed model can be inverted and optimal parameters identified means that distributed hydrologic modeling is possible and not necessarily over-determined because of the number of degrees of freedom. As with the first and second editions, scientific principles contained herein relate to capturing essential scales of spatial variability, information content, and how to apply a fully distributed physics-based model. This includes application to small urban catchments and large river basins, with time-frames spanning from short-event scales lasting an hour or less, to long-term climate change projections across periods of years, decades, or centuries. The result of this approach is intended to guide hydrologists in the pursuit of more reliable hydrologic prediction.

References

Clark, M.P., B. Nijssen, J.D. Lundquist, D. Kavetski, D.E. Rupp, R.A. Woods, J.E. Freer, E.D. Gutmann, A.W. Wood, L.D. Brekke, and J.R. Arnold. 2015a. A unified approach for process-based hydrologic modeling: 1. Modeling concept. *Water Resources Research* 51(4): 2498–2514.

Clark, M.P., B. Nijssen, J.D. Lundquist, D. Kavetski, D.E. Rupp, R.A. Woods, J.E. Freer, E.D. Gutmann, A.W. Wood, D.J. Gochis, and R.M. Rasmussen. 2015b. A unified approach for process-based hydrologic modeling: 2. Model implementation and case studies. *Water Resources Research* 51(4): 2515–2542.

Downer, C.W., and F.L. Ogden. 2004. GSSHA: a model for simulating diverse streamflow generating processes. *Journal of Hydrologic Engineering* 9(3): 161–174.

Farajalla, N.S., and B.E. Vieux. 1995. Capturing the essential spatial variability in distributed hydrologic modeling: Infiltration parameters. *Journal of Hydrology Processes* 9(1): 55–68.

Gochis, D.J., W. Yu, D.N. Yates. 2014. The WRF-Hydro model technical description and user's guide, version 2.0. NCAR Technical Document.

Green, W.H., and G.A. Ampt. 1911. Studies in soil physics I: The flow of air and water through soils. *Journal of Agricultural Science* 4: 1–24.

Julien, P.Y. and B. Saghafian. 1991. *CASC2D User's Manual*. Civil Engineering Report. Fort Collins, Colorado: Dept. of Civil Engineering, Colorado State University.

Julien, P.Y., B. Saghafian, and F.L. Ogden. 1995. Raster-based hydrological modeling of spatially-varied surface runoff. *Water Resources Bulletin, AWRA* 31(3): 523–536.

Le Dimet, F.X., H.E. Ngodock, and M. Navon. 1996. *Sensitivity analysis in variational data assimilation*. Santa Fe, New Mexico, USA: Siam Meeting in Automatic differentiation.

Le Dimet, F.-X., W. Castaings, P. Ngnepieba and B. Vieux. 2009. Data assimilation in hydrology: variational approach. In *Data assimilation for atmospheric, oceanic and hydrologic applications*. Berlin Heidelberg: Springer, pp. 367–405.

Loague, K.M. 1988. Impact of rainfall and soil hydraulic property information on runoff predictions at the hillslope scale. *Water Resources Research* 24(9): 1501–1510.

Loague, K., C.S. Heppner, B.A. Ebel, and J.E. VanderKwaak. 2010. The quixotic search for a comprehensive understanding of hydrologic response at the surface: Horton, Dunne, Dunton, and the role of concept-development simulation. *Hydrological Processes* 24(17): 2499–2505.

Moussa, R., and C. Bocquillon. 1996. Criteria for the choice of flood-routing methods in natural channels. *Journal of Hydrology* 186: 1–30.

National Weather Service (NWS). 2013. Exceedance Probability Analysis for the Colorado Flood Event, 9–16 September 2013 Hydrometeorological Design Studies Center National Weather Service National Oceanic and Atmospheric Administration 1325 East-West Highway, Silver Spring, MD.

Ponce, V.M., R.N. Li, and D.B. Simons. 1978. Applicability of kinematic and diffusion models. *Journal of the Hydraulics Division American Society of Civil Engineers* 104(3): 353–360.

Sene, Kevin. 2012. *Flash floods: forecasting and warning*. Springer Science & Business Media.

Smith, M.B., D.J. Seo, V.I. Koren, S. Reed, Z. Zhang, Q.Y. Duan, S. Cong, F. Moreda, and R. Anderson. 2004. The distributed model intercomparison project (DMIP): Motivation and experiment design. *Journal of Hydrology* 298(1–4): 4–26.

Tachikawa, Y., M. Shiiba, and T. Takasao. 2007. development of a basin geomorphic information system using a T-Dem data structure. *JAWRA Journal of the American Water Resources Association* 30(1): 9–17.

Verhoest, N.E., and P.A. Troch. 2000. Some analytical solutions of the linearized Boussinesq equation with recharge for a sloping aquifer. *Water Resources Research* 36(3): 793–800.

Vieux, B.E. 1988, *Finite element analysis of hydrologic response areas using geographic information systems*. Department of Agricultural Engineering, Michigan State University. A dissertation submitted in partial fulfillment of the degree of Doctor of Philosophy.

Vieux, B.E., V.F. Bralts, L.J. Segerlind, and R.B. Wallace. 1990. Finite element watershed modeling: one-dimensional elements. *Journal of Water Resources Planning and Management* 116(6): 803–819.

Vieux, B.E. 1991. Geographic information systems and non-point source water quality modeling. *Journal of Hydrology Processes* 5: 110–123.

Vieux, B.E. 1993. DEM aggregation and smoothing effects on surface runoff modeling. *ASCE Journal of Computing in Civil Engineering, Special Issue on Geographic Information Analysis*, 7(3): 310–338.

Vieux, B.E., and N.S. Farajalla. 1994. Capturing the essential spatial variability in distributed hydrologic modeling: Hydraulic roughness. *Journal of Hydrology Processes* 8(3): 221–236.

Vieux, B.E., and N. Gauer. 1994. Finite element modeling of storm water runoff using GRASS GIS. *Microcomputers in Civil Engineering* 9(4): 263–270.

Vieux, B.E. and J.E. Vieux. 2002. V*flo*®: A Real-time distributed hydrologic model. In *Proceedings of the 2nd Federal Interagency Hydrologic Modeling Conference*. Las Vegas, Nevada July 28–August 1, 2002.

Vivoni, E., V. Ivanov, R. Bras, and D. Entekhabi. 2004. Generation of triangulated irregular networks based on hydrological similarity. *Journal of Hydrologic Engineering* 9(4): 288–302.

White, L.W., B.E. Vieux, and D. Armand. 2002. Surface flow model: Inverse problems and predictions. *Journal Advances in Water Resources* 25(3): 317–324.

White, L., B. Vieux, D. Armand, and F.-X. Le Dimet. 2003. Estimation of optimal parameters for a surface hydrology model. *Journal Advances in Water Resources* 26(3): 337–348.

Chapter 2
Geospatial Data for Hydrology

Data Sources and Structure

Abstract A key characteristic of distributed modeling is the spatially variable representation of the watershed in terms of topography, vegetative, or land use/cover, soils and impervious areas and the derivative model parameters that govern the hydrologic processes of infiltration, evapotranspiration, and runoff. Geospatial data exist that can be harnessed for model setup. Digital representation of topography, soils, land use/cover, and precipitation may be accomplished using widely available or special purpose GIS data sets. Each GIS data source has a characteristic data structure, which has implications for the hydrologic model. Major data structures are *raster* and *vector*. Raster data structures are characteristic of remotely sensed data with a single value representing a grid cell. Points, polygons, and lines are referred to generally as *vector* data. Multiple attributes can be associated with a point, line, area, or grid cell. Some data sources capture characteristics of the data in terms of measurement scale or sample volume. A rain gauge essentially measures rainfall at a point, whereas radar, satellites and other remote sensing techniques typically map the spatial variability over large geographic areas at resolutions ranging from meters to kilometers. Source data structures can have important consequences on the derived parameter and, therefore, model performance. Even after considerable processing, hydrologic parameters can continue to have some vestige of the original data structure, which is termed an *artifact*. This chapter addresses geospatial data structure, projection, scale, dimensionality, and source data for hydrologic applications.

2.1 Introduction

When setting up a distributed hydrologic model, a variety of data structures may be employed. Topography, for example, may be represented by a series of point elevation measurements or may exist as contour lines, grids, or triangular facets composing a triangular irregular network (TIN). Rainfall may be represented by a point when taken from rain gauges but exist as grids when measured by radar. Infiltration rates derived from soil maps are generalized over the polygon describing

© Springer Science+Business Media Dordrecht 2016   21
B.E. Vieux, *Distributed Hydrologic Modeling Using GIS*,
Water Science and Technology Library 74, DOI 10.1007/978-94-024-0930-7_2

the soil mapping unit. Even though there exists variability within such mapping units, its properties are considered constant within the polygon enclosing a single soil type. Resulting infiltration modeled using this data source will likely show artifacts of the original soil map polygons such as abrupt changes at the boundary of adjacent soil mapping units. Land use/cover may be used to develop evapotranspiration rates or estimates of hydraulic roughness from polygonal areas or from a raster array of remotely sensed surrogate measures. In any case, the spatial variability of the parameters may be affected by the data structure of the source.

These examples illustrate two important points. First, in most cases a data source may be either a direct measure of the physical characteristic or an indirect (surrogate) measure requiring conversion or interpretation. Second, because of how the data are measured, each source has a characteristic structure including spatial and temporal dimensions as well as geometric character (points, lines, polygons, rasters, or polar arrays of radar data). Because the model may not be expecting data in one form or another, transformation is often necessary from one data structure to another. This necessity often arises because hydrologic processes/parameters are not directly observable at the scale expected by the model, or because the spatial or temporal scale of the measured parameter differs from that of the model. This issue can require transformations from one projection to another, from one structure to another (e.g., contours to TINs), interpolation of point values, or surface generation. Because the GIS data form the basis for numerical algorithms, hydrologic modeling requires more complex GIS analysis than simple geographical modeling using maps.

Typically, distributed hydrologic modeling divides a watershed or region into computational elements. Given that the numerical algorithms used to solve conservation of mass and momentum equations in hydrology may divide the domain into discrete elements, a parameter may be assumed constant within the computational element. At the sub-grid scale, parameter variation is present and may be represented in the model as statistical distribution known as sub-grid parameterization. Hydraulic roughness may be measured at a point by relating flow depth and velocity. In a distributed model, the roughness is assigned to a grid cell based on the dominant land use/cover classification. In most applications, the model computational element will not conform exactly to the measured parameter. Conformance of the data structure in the GIS map to the spatial pattern or scale of the process is a basic issue in GIS analysis for hydrology. Transformation from one data structure to another must be dealt with for effective use of GIS data in hydrology. The components of data structure are treated below.

2.2   Map Scale and Spatial Detail

A map of hydrography can be shown at any scale within a GIS. Once data are digitized and represented electronically in a GIS, resolution finer than that at which it was compiled is lost. A small-scale map is one in which features appear small, have few details, and cover large areas. An example of a small-scale map is one with a scale of 1:1,000,000. Conversely, large-scale maps have features that appear large and cover small areas. An example of a large-scale map is one with a scale of 1:2,000. A map compiled at 1:1,000,000 can easily be displayed in a GIS at a 1:2,000 scale, giving the false impression that the map contains more information than it really does.

Because GIS provides the ability to easily display data at any scale, we must distinguish between the compilation or native scale and the user-selected scale. The scale in a geographic context is large if features appear large (e.g., 1:1,000) and conversely small if features appear small (1:1,000,000). The scale and resolution at which the data are collected or measured is termed the **native** scale or resolution. If the spot or point elevations are surveyed in the field on a grid of 100 m, this is its native resolution. Once contours are interpolated between the points and plotted on a paper map at a scale of, for example, 1:25,000, we have introduced a scale to the data. Once the paper map is digitized, there will be little more information contained at a scale smaller than 1:25,000; enlarging to 1:1,000 would make little sense because detail was lost or never captured at 1:25,000. The importance to hydrologic modeling is that variations in landform, stream channels, or watershed boundaries may not be adequately captured at resolutions that are too coarse or at small scales greater than 1:100,000. Simply changing the scale of a map compiled at 1:25,000–1:1,000 could be misleading if small variations were lost or never captured when the map was compiled at the smaller scale of 1:25,000.

The hydrologist must decide what scale will best represent hydrologic processes. If micro-topography at the scale of rills or small rivulets controls the rate of erosion and sediment transport, then small-scale maps (e.g., 1:1,000,000) will contain little information relevant to the modeling of the process. However, such a map may contain sufficient detail for modeling the river basin hydrologic response.

2.3   Georeferenced Coordinate Systems

In geodesy, a georeferenced coordinate system is based on a vertical and horizontal *datum*. The classical datum is defined by five elements, which give the position of the origin (two elements), the orientation of the network (one element), and the parameters of a referenced ellipsoid (two elements). The World Geodetic System (WGS) is a geocentric system that provides a basic reference frame and geometric figure for the Earth, models the Earth gravimetrically and provides the means for relating positions on various data to an Earth-centered, Earth-fixed coordinate system. Even if two maps are in the same coordinate system, discrepancies may still be apparent due to a different datum or scale used to compile each map. This problem is common when a map is compiled with an older datum and then used with data compiled with an updated datum. The usual remedy is to adjust the older datum to bring it into alignment with the revised datum. Conversion routines exist to transform spatial data from one datum to another.

Correction from one datum to another will not remove differences caused by compilation scale. If the aerial photography is collected at 1:25,000 but the hydrography was compiled at a smaller scale, then the streams will not line up with the photography. Registration can be a problem when using generally available geospatial data sets that have been compiled at disparate scales. A similar effect of misregistration may be observed when combining vector hydrography and raster elevation data because of differences in the scale at which the respective geospatial data were compiled.

Digital aerial photography is available for many parts of the US either through government-sponsored acquisition or on a project basis. A GIS should be able to properly overlay geospatial data that exist in two different georeferenced coordinate systems. For example, aerial photography that is georeferenced is called orthophotography or orthophoto. Because the aerial photo was adjusted through geometric transformations, it can be overlaid with other maps such as stream channels. An example of how georeferencing can produce obvious errors is seen in Fig. 2.1 where two streams are overlaid on top of a digital orthophoto. The orthophoto and stream shown in black were compiled in North American Datum 83 (NAD83), whereas the second stream (shown in white) is apparently displaced to the south and east, because it was compiled in the NAD27 datum. The NAD83 stream (red) matches well with stream channel features in the orthophoto because their data are consistent.

Fig. 2.1 Mismatched datum effects on location of hydrographic features (*red stream*, NAD83; *white stream*, NAD27)

The NAD27 stream (shown in white) is inconsistent with the orthophoto datum and like differs due to channel morphology since the original map compilation.

Georeferenced coordinate systems are developed to consistently map features on the Earth's surface. Each point on the Earth's surface may be located by a pair of latitude and longitude. Because the Earth is an oblate spheroid, the distance between two points on that surface depends on the assumed radius of the Earth at the particular location. Both spherical and ellipsoidal definitions of the Earth exist. The terms *spheroid* and *ellipsoid* refer to the definition of the dimensions of the Earth in terms of the radii along the equator and along a line joining the poles. As better definitions of the spheroid are obtained, the spheroid has been updated compared to historically older definitions that arose before the advent of satellite measurements such as the Clark spheroid developed in 1866. Recent spheroids derived from satellite measurements include the GRS1980 and WGS84. Periodically, the vertical or horizontal datum is updated with the aid of geodetic surveys or GPS measurements.

2.4 Map Projections

A variety of map projections exist that transform three-dimensional spherical/ellipsoidal coordinates, expressed in latitude and longitude, to two-dimensional planar coordinates. Together with a geoid and datum definition, the equations are called a *projection*. Depending on the source of the GIS data, one may encounter a variety of map projections. When distances or areas are needed from geospatial data, the data are almost always projected from latitude and longitude into a two-dimensional plane.

All projections introduce distortion, because the projection transforms positions located on a three-dimensional surface, i.e., spheroid, to a position located on a two-dimensional surface, called the projected surface. There are three main types of projections. *Conformal* projections by definition maintain local angles between the original decimal degree and the projected reference system. This means that if two lines intersect each other at an angle of 30° on the spheroid, then in a conformal projection, the angle is maintained on the projected surface only if the projection is *equal distance*. The stereographic projection is conformal but not equal area or equal distance. Because hydrology is often concerned with distances and areas, map projections that preserve these quantities find broadest usage.

The usefulness of maps for navigation has made geographic projections an important part of human history. Figure 2.2 shows the countries of the world projected onto a plane tangent to the North Pole using the stereographic projection. While this polar form was probably known by the Egyptians, the first Greek to use it was Hipparcus in the second century. In 1613, Francois d'Aiguillon was the first to name it *stereographic* (Snyder 1987).

While this projection has long been used for navigational purposes, it has been used more recently for hydrologic purposes. The US National Weather Service

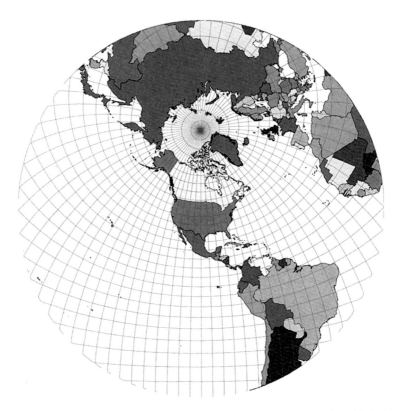

Fig. 2.2 Stereographic projection of countries together with *lines* of longitude and latitude

(NWS) uses it to map radar estimates of rainfall on a national grid called HRAP. Figure 2.3 shows the basic idea of the stereographic projection. The choice of map plane latitude, termed *reference latitude*, is a projection parameter that depends on the location and extent of features to be mapped. The distance between A′ and B′ is less on the map plane at 60 °N latitude than the distance between A″ and B″ at 90 °N latitude. Changes in distance on a given plane can constitute a distortion.

Parameters of a projection and the type of projection are important choices since the accuracy of mapped features depends on the selection. The spheroid may be represented with a single radius in the case of spherical definitions versus ellipsoidal definitions that contain a major and minor radius. Countries develop their own geodetic coordinate system. NAD 83 defined above is the horizontal datum on which many projections are based in North America. In other countries, a datum and coordinate system may also be defined, for example the Korean 1985 projection that consists in the Transverse Mercator projection (discussed below) and the Bessel 1941 spheroid with an equatorial radius of 6,377,397.155 m and a polar radius of 6,356,079.0, resulting in a flattening ratio of 299.1528128.

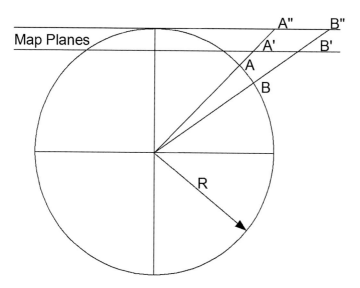

Fig. 2.3 Stereographic projection method for transforming the geographic location of *points A* and *B* to a map plane

In choosing a projection, we seek to minimize the distortion in angles, areas, or distances, depending on which aspect is more important. There is no projection that maintains all three characteristics because a geographic feature in three dimensions (elevation, latitude, and longitude) to two dimensions on a planar map always introduces some distortion. While mathematically there is no projection that simultaneously preserves local angles (conformal), or preserves area or distance, we can preserve two of the three quantities. For example, the Universal Transverse Mercator (UTM) projection is designed to be *conformal* and *equal area* though not *equal distance*. Figure 2.4 shows a transverse developable surface tangent along a meridian of longitude. To minimize the distortion in distance, the UTM projection divides the Earth's surface into 60 zones. In mid-latitudes around the world, identical projections are made in each zone of 6° longitude (360°/60). This means that the coordinates in the projected surface uniquely describe a point only within the zone. The projected coordinates are in meters with the *x*-coordinate (east-west) of 500,000 m being assigned to the central meridian of each zone. It is not enough to simply say that a particular point is located at $x = 500,000$ and $y = 2,000,000$ m, because this does not uniquely define the location in the UTM projection, we must also specify the zone.

Depending on the projection, distortions in either the east-west direction or the north-south direction may be minimized but we cannot have both. If the geographic feature spans more distance in the north-south direction, a projection that minimizes distortion in this direction is often used. The direction of least distortion is determined by the *developable surface* used to transform the coordinates on the spheroid

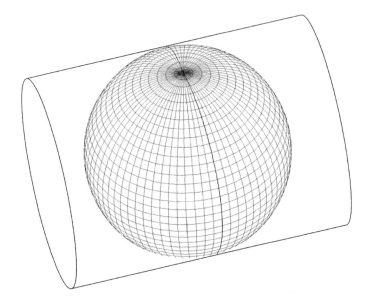

Fig. 2.4 Transverse cylinder developable surface used in UTM projection

to a two-dimensional surface. Typical developable surfaces are cylinders or cones. The orientation of the axis of the developable surface with respect to the Earth's axis determines whether the projection is termed transverse or oblique. A *transverse projection* would orient a cylinder whose axis is at right angles to the Earth's axis and is tangent to the Earth's surface along some meridian. The cylinder is then "unwrapped" to produce a two-dimensional surface with Cartesian coordinates. In an oblique projection, the axis of the developable surface and the Earth's axis form an oblique angle. Distortion in distance is minimized in the UTM projection by making this cylinder tangent at the central meridian of each zone and then unwrapping it to produce the projected map.

Choosing the appropriate projection for the spatial extent of a hydrologic feature is important. Figure 2.5 shows the Arkansas-Red-White River basins and subbasins along with state boundaries. The graticule at 6° intervals in longitude corresponds to the UTM zones. Zones 13, 14, and 15 are indicated at the bottom of Fig. 2.5.

The watershed boundaries shown are in decimal degrees of latitude and longitude and not projected. A watershed that crosses a UTM zone cannot be mapped because the x-coordinate is nonunique. The origin starts over at 500,000 m at the center of each zone. The UTM projection is an acceptable projection for hydrologic analysis of limited spatial extent provided the area of interest does not intersect a UTM zone boundary. Many of the subbasins shown in Fig. 2.5 above could effectively be mapped and modeled using UTM coordinates, whereas the entire Arkansas-Red-White basin could not because it intersects two or more UTM zones. Selecting the appropriate projection and datum for hydrologic analysis depends on

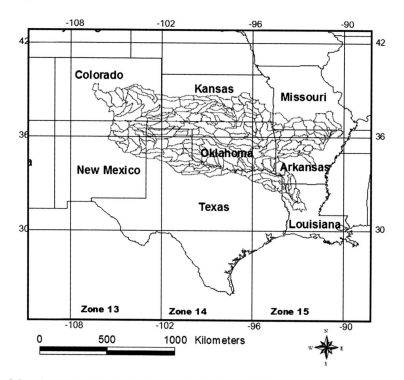

Fig. 2.5 Arkansas-*Red-White* River basin and UTM zones 13, 14 and 15

the extent of the watershed or region. If the distance spanned is large, then projection issues take on added importance. In small watersheds, the issue of projection is not as important because errors or distortions are on the order of parts per million and do not generally have a significant impact on the hydrologic simulation. In fact, for small watersheds we can use an assumed coordinate system that relates features to each other without regard to position on the Earth's surface. The following section deals with general concepts of geographic modeling along with the major types of data representation with specific examples related to hydrology.

2.5 Data Representation

2.5.1 Metadata

Knowing the origin, lineage, and other aspects of the data you are using is essential to understanding its limitations and getting the most usefulness from the data. This information is commonly referred to as data about the data, or *metadata*. For example, the fact that an elevation data set is recorded to the nearest meter could be discovered from the metadata. If the topography controlling the hydrologic process is controlled by surface features, e.g., rills and small drainage channels that physically have scales less than 1 m in elevation or spatial extent, then digital representation of the surface may not be useful in modeling the hydrologic process. An example of metadata documentation for GIS data is provided by the Federal Geographic Data Committee (FGDC) with content standards for digital geospatial metadata (FGDC 1998). The FGDC-compliant metadata files contain detailed descriptions of the data sets and narrative sections describing the procedures used to produce the data sets in digital form. Metadata can be indispensable for resolving problems with data. Understanding the origin of the data, the datum and scale at which it was originally compiled and the projection parameters are prerequisites for effective GIS analysis. The following section describes the major methods for representing topography and its use in automatic delineation of watershed areas. Types of surface representation can be extended to other attributes besides elevation such as rainfall.

2.5.2   Topographic Representation

A *digital elevation model* (DEM) consists in an ordered array of numbers representing the spatial distribution of elevations above some arbitrary datum in a landscape. It may consist in elevations sampled at discrete points or the average elevation over a specified segment of the landscape, although in most cases it is the former. DEMs are a subset of *digital terrain models* (DTMs), which can be defined as ordered arrays of numbers that represent the spatial distribution of terrain attributes, not just elevation. Over 20 topographic attributes can be used to describe landform. The slope and direction of principal land surface gradients are important and widely used terrain model attributes. Other primary topographic attributes include the specific catchment area and altitude. The terrain and land surface slope plays an important role in hydrologic processes, especially runoff. Reliable estimation of topographic parameters reflecting terrain geometry is necessary for geomorphological, hydrologic, and ecological studies, because terrain controls runoff, erosion, and sedimentation.

When choosing the particular method of representing the surface, it is important to consider the end use. The ideal structure for a DEM may be different if it is used as a structure for a distributed hydrologic model than if it is used to determine the topographic attributes of the landscape. There are three principal ways of structuring a network of elevation data for its acquisition and analysis

1. Contour
2. Raster
3. Triangular irregular network

With this introduction to the three basic types of surface representation, we now turn in more detail to each of the three.

2.5.2.1   Contour

A *contour* is an imaginary line on a surface showing the location of an equal level or value. For example, an isohyet is a line of equal rainfall accumulation. Representation of a surface using contours shows gradients and relative minima and maxima. The interval of the contour is important, particularly when deriving parameters. The difference between one level contour and another is referred to as *quantization*. The hydrologic parameter may be quantized at different intervals depending on the variability of the process and the scale at which the hydrologic process is controlled. Urban applications often utilize contours produced at 2-ft (0.61 m) intervals, whereas, USGS typically produces contours at 10-ft (3.048 m) intervals for general mapping purposes.

Contour-based methods of representing the land surface elevations or other attributes have important advantages for hydrologic modeling because the structure of their elemental areas is based on the way in which water flows over the land surface (Moore et al. 1991). Lines orthogonal to the contours are streamlines, so the equations describing the flow of water can be reduced to a series of coupled one-dimensional equations. Many DEMs are derived from topographic maps, so their accuracy can never be greater than the original source of data. For example, the most accurate DEMs produced by the United States Geological Survey (USGS) are generated by linear interpolation of digitized contour maps and have a maximum root mean square error (RMSE) of one-half contour interval and an absolute error of no greater than two contour intervals in magnitude.

Developing contours in hydrologic applications carries more significance than merely representing the topography. Under certain assumptions, it is reasonable to assume that the contours of the land surface control the direction of flow. Thus, surface generation schemes that can extract contour lines and orthogonal streamlines at the same time from the elevation data have advantages of efficiency and consistency.

2.5.2.2   Raster

The raster data structure is perhaps one of the more familiar data structures in hydrology. Many types of data, especially remotely sensed information, are often measured and stored in raster format. The term *raster* derives from the technology developed for television in which an image is composed of an array of picture elements called pixels. This array or raster of pixels is also a useful format for representing geographical data, particularly remotely sensed data, which in its native format is a raster of pixels. Raster data are also referred to as grids. Because of the vast quantities of elevation data that are in raster format, it is commonly used for watershed delineation, deriving slope, and extracting drainage networks. Other surface attributes such as gradient and aspect may be derived from the DEM and stored in a *digital terrain model* (DTM). The term DTED stands for *digital terrain elevation data* to distinguish elevation data from other types of DTM attributes.

Raster DEMs are one of the most widely used data structures because of the ease with which computer algorithms are implemented. However, they do have several disadvantages:

- They cannot easily handle abrupt changes in elevation.
- The size of grid mesh affects the results obtained and the computational efficiency.
- The computed upslope flow paths used in hydrologic analyses tend to zig-zag and are therefore somewhat unrealistic.
- The delineation of catchment areas may be imprecise in flat areas.

Capturing surface elevation information in a digital form suitable to input into a computer involves sampling x, y, z (easting, northing, and elevation) points from a model representing the surface, such as a contour map, stereo-photographs, or other images. DEMs may be sampled using a variety of techniques. Manual sampling of DEMs involves overlaying a grid onto a topographic map and manually coding the elevation values directly into each cell. However, this is a very tedious and time-consuming operation suitable only for small areas. Alternatively, elevation data may be sampled by direct quantitative photogrammetric measurement from aerial photographs on an analytical stereo-plotter. More commonly, digital elevation data are sampled from contour maps using a digitizing table that translates the x, y, and z data values into digital files. Equipment for automatically scanning line maps has also been developed based on either laser-driven line-following devices or a raster scanning device, such as the drum scanner. However, automatic systems still require an operator to nominate the elevation values for contour data caused by poor line work, the intrusion of non-contour lines across the contour line being automatically scanned, or other inconsistencies. DEMs may be derived from overlapping remotely sensed digital data using automatic stereo-correlation techniques, thereby permitting the fast and accurate derivation of DEMs. With the increasing spatial accuracy of remotely sensed data, future DEMs will have increasingly higher accuracy. Van Zyl (2001) described the US Space Shuttle mission that mapped 80 % of the populated Earth surface to 30 m resolution. Because of *holes* in this data set, an improved data set was produced and made available by Jarvis et al. (2008).

The major disadvantage of grid DEMs is the large amount of data redundancy in areas of uniform terrain and the subsequent inability to change grid sizes to reflect areas of different complexity of relief (Burrough 1986). However, various techniques of data compaction have been proposed to reduce the severity of this problem, including quadtrees, freeman chaincodes, run-length codes, and block codes used to compress the raster data structure. Advantages of a regularly gridded

DEM are its easy integration with raster databases and remotely sensed digital data, the smoother, more natural appearance of contour maps and derived terrain features maps and the ability to change the scale of the grid cells rapidly.

2.5.2.3   Triangular Irregular Network

A *triangular irregular network* (TIN) is an irregular network of triangles representing a surface as a set of non-overlapping contiguous triangular facets of irregular sizes and shapes. TINs are more efficient at representing the surface than the uniformly dense raster representation. TINs have become increasingly popular because of their efficiency in storing data and their simple data structure for accommodating irregularly spaced elevation data. Advantages have also been found when TIN models are used in inter-visibility analysis on topographic surfaces, extraction of hydrologic terrain features, and other applications.

A TIN has several distinct advantages over contour and raster representations of surfaces. The primary advantage is that the size of each triangle may be varied such that broad flat areas are covered with a few large triangles, while highly variable or steeply sloping areas are covered with many smaller triangles. This provides some efficiency over raster data structures since the element may vary in size according to the variability of the surface. Given the advantages of TINs in representing data requiring variable resolution, we will examine the features of and methods for generating the TINs.

A TIN approximates a terrain surface by a set of triangular facets. Each triangle is defined by three edges and each edge is bound by two vertices. Most TIN models assume planar triangular facets for the purpose of simplifying interpolation or contouring. Vertices in TINs describe nodal terrain features, e.g., peaks, pits, or passes, while edges depict linear terrain features, e.g., break, ridge or channel lines. Building TINs from grid DEMs therefore involves some procedures for efficiently selecting the locations of TIN vertices for nodal terrain features or TIN edges for linear terrain features. Grid DEMs are widely available at a relatively low cost. Because of this increasing availability, the need for an efficient method to extract critical elevation points from grid DEMs to form TINs has increased. Some caution should be exercised to ensure that critical features are not lost in the conversion process. Figure 2.6 is a TIN created from a grid DEM. Notice the larger triangles representing flat areas and the smaller, more numerous, triangles in areas where there is more topographic relief. Using triangles of various sizes, where needed, reduces computer storage compared with raster formats.

Methods for constructing a TIN from a grid DEM were investigated by Lee (1991). In general, methods for creating TINs consist in selecting critical points from grid DEMs. Nonessential grid points are discarded in favor of representing the surface with fewer points linked by triangular facets. These conversion methods may be classified into four categories: (1) skeleton, (2) filter, (3) hierarchy, and (4) heuristic method. Each of these methods has its advantages and disadvantages,

Fig. 2.6 TIN elevation model derived from a gridded LiDAR DEM

different in how they assess *point* importance and in their *stopping rules*, i.e., when to stop eliminating elevation points. The common property of all the methods is that the solution sets depend on some predetermined parameters; these are several tolerances in the skeleton method and either a tolerance or a prescribed number of output points in the other three methods.

The objective of any method of converting DEMs to TINs is to extract a set of irregularly spaced elevation points that are as few as possible while at the same time providing as much information as practical about topographic structures. Unfortunately, it is impossible to achieve both goals concurrently and difficult to balance the two effects. Surface geometry that is highly variable is best represented by a range of smaller to larger TINs where needed. Compared with raster, an advantage of the TIN format is that small variations in terrain such as hillslopes or road embankments can be modeled along with broad-scale features such as the floodplain along a river with triangular facets of varying sizes. The importance of digital elevation data and derivative products for distributed hydrologic modeling cannot be overstated. Cell resolution effects are discussed in Chap. 4, which deals with information content and spatial variability. Chapter 6 addresses in detail aspects of drainage networks derived from DEMs. Methods for automatic delineation of watershed boundaries from DEMs are discussed in the Sect. 2.6.

2.6   Watershed Delineation

Raster or TIN are the primary data structures used in the delineation of watershed boundaries by automated methods. Defining the watershed and the drainage network forms the basic framework for applying both lumped and distributed hydrologic models. Moore et al. (1991) discussed the major data structures for watershed delineation ranging from grid, TIN, and contour methods. Figures 2.7 and 2.8 show the Illinois River Basin delineated to just below Lake Tenkiller in Eastern Oklahoma. The watershed area is 4,211 km^2, delineated at a DEM resolution of 60 m (Fig. 2.7) and 1080 m (Fig. 2.8), respectively. At 60 m the number of cells is 1,169,811, while the delineated watershed at 1,080 m resolution has much fewer with only 3,690 cells. While the larger resolution reduces computer storage, sampling errors increase. The variability of the surface may not be adequately captured at coarse resolution, resulting in difficulties in automatic watershed delineation. The two delineated watersheds look similar but have slightly different watershed boundaries. The difference in boundary shape can be more severe where flatter slopes are not adequately resolved by the vertical resolution, in this case to the nearest 1 m.

Depressions may be natural features or simply a result of sampling an irregular surface with a regular sampling interval, i.e., grid resolution. The hydrologic significance of depressions depends on the type of landscape represented by the DEM.

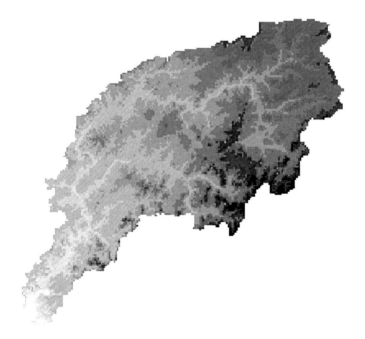

Fig. 2.7 Digital elevation map of the Illinois River Basin with 60 m resolution

Fig. 2.8 As in Fig. 2.7 but
delineated at 1,080 m

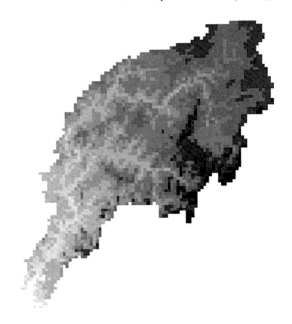

In some areas, such as the prairie pothole region in the Upper Midwest (North Dakota) of the United States or in parts of the Sahel in Africa, surface depressions dominate and control the hydrologic processes. In areas with coordinated drainage, depressions are an artifact of the sampling and generation schemes used to produce the DEM. We should distinguish between real depressions and those that are artifacts introduced by the sampling scheme and data structure.

2.6.1 Algorithms for Delineating Watersheds

O'Callaghan and Mark (1984), Peucker and Douglas (1975) and Jenson (1987) proposed algorithms to produce depressionless DEMs from regularly spaced grid elevation data. Numerical filling of depressions, whether from artifacts or natural depressions, facilitates the automatic delineation of watersheds. By smoothing a DEM these methods are capable of determining flow paths iteratively, especially where there is more than one possible receiving cell and where flow must be routed across flat areas. The three main methods examined by Skidmore (1990) for calculating ridge and gully position in the terrain are

1. Peucker and Douglas algorithm—Maps ridges and valleys using a simple moving-window algorithm. The cell with the lowest elevation in a two-by-two moving window is flagged. Any unflagged cells remaining after the algorithm has passed over the DEM represent ridges. Similarly, the highest cell in the

window is flagged, with any unflagged cells in the DEM corresponding to valley
lines.

2. O'Callaghan and Mark algorithm—Forms a DEM based on quantifying the
   drainage accumulation at each cell in the DEM. Cells that have a drainage
   accumulation above a user-specific threshold are considered to be on a drainage
   channel. Ridges are defined as cells with no drainage accumulation.

3. Band algorithm—Band (1986) proposed a method for identifying streamlines
   from a DEM, Rosenfeld and Kak (1982) thinning algorithm. The upstream and
   downstream nodes on each stream fragment are then flagged. Each downstream
   node is "drained" along the line of maximum descent until it connects with
   another streamline. The streams are again thinned to the final, one-cell wide, line
   representation of the stream network.

More recent research has improved upon these models and many GIS modules
exist for processing DEMs for purposes of extracting stream networks and delineating watersheds as discussed in Chap. 7. The decision on which algorithm to use
depends largely on the resolution of indeterminate flow directions caused by flat
slopes.

2.6.2   Problems with Flat Areas

Truly flat landscapes, or zero slope, seldom occur in nature. Yet, when a landscape
is represented by a DEM, areas of low relief can translate into flat surfaces. This
flatness may also be a result of vertical quantization (precision) of the elevation
data. This occurs when the topographic variation is less than 1 m yet the elevation
data are reported with a precision to the nearest meter. Flat surfaces are typically the
result of inadequate vertical DEM resolution, which can be further worsened by a
lack of horizontal resolution. Such flat surfaces are also generated when depressions
in the digital landscape are removed by raising the elevations within the depressions
to the level of their lowest flow.

A variety of methods has been proposed to address the problem of drainage
analysis over flat surfaces. Methods range from simple DEM smoothing to arbitrary
flow direction assignment. However, these methods have limitations. DEM
smoothing introduces loss of information to the already approximate digital elevations, while arbitrary flow direction assignment can produce patterns that reflect
the underlying assignment scheme, which are not necessarily realistic or topographically consistent. Given these limitations, the application of automated DEM
processing is often restricted to landscapes with well-defined topographic features
that can be resolved and represented by the DEM. Improved drainage identification
is needed over flat surfaces to extend the capabilities and usefulness of automated
DEM processing for drainage analysis.

Garbrecht and Martz (1997) presented an approach that produced more realistic
and topographically consistent drainage patterns than those provided by earlier

methods. The algorithm increments cell elevations of the flat surface to include information on the terrain configuration surrounding the flat surface. As a result, two independent gradients are imposed on the flat surface: one away from the higher terrain into the flat surface and the other out of the flat surface towards lower terrain. The linear combination of both gradients, with localized corrections, is sufficient to identify the drainage pattern while at the same time satisfying all boundary conditions of the flat surface. Imposed gradients lead to more realistic and topographically consistent drainage over flat surfaces. The shape of the flat surface, the number of outlets on its edge, and the complexity of the surrounding topography apparently do not restrict the proposed approach. A comparison with the drainage pattern of an established method that displays the "parallel flow" problem shows significant improvements in producing realistic drainage patterns.

One of the most satisfactory methods for assigning drainage directions on flat areas is that of Jenson and Dominique (1988). The Jenson and Domingue (JD) algorithm is useful over most of the DEM but does not produce satisfactory results in areas of drainage lines because it causes these lines to be parallel. The JD algorithm assigns drainage directions to flat areas in valleys and drainage lines such that the flow is concentrated into single lines and it uses the JD method over the rest of the DEM where less convergent flow is more realistic. Automated valley and drainage network delineation seeks to produce a fully connected drainage network of single cell width because this is what is required for applications such as hydrologic modeling. No automatically delineated drainage network is likely to be very accurate in flat areas, because drainage directions across these areas are not assigned using information directly held in the DEM.

One method for enforcing flow direction is to use an auxiliary map to restrict drainage direction where a mapped stream channel exists. Turcotte et al. (2001) described a method for incorporating a river and lake network into the delineation process that yields a more satisfying result where large flat areas or lakes form a part of the drainage network. If an accurate river or stream vector map is available, it may be used to *burn in* the elevations, forcing the drainage network to coincide with the vector map depicting the desired drainage network. By burning in, i.e., artificially lowering the DEM at the location of mapped streams, the correct location of the automatically delineated watershed and corresponding stream network is preserved. Watersheds and stream networks delineated from a DEM become a type of data structure for organizing lumped and distributed hydrologic model computations. The following section deals with another type of map representation useful in soil mapping or other thematic representations.

2.7 Soil Classification

An important source of hydrologic modeling parameters is a map soils for simulating infiltration. Mapping soils usually involves delineating soil types that have identifiable characteristics. The delineation is based on many factors germane to soil science, such as geomorphologic origin and conditions under which the soil formed, e.g., grassland or forest. Regardless of the purpose or method of delineation, there will be a range of soil properties within each mapping unit. This variation may stem from inclusions of other soil types too small to map and from natural variability.

The primary hydrologic interest in soil maps is the modeling of infiltration as a function of soil properties associated with mapped polygons. Adequate measurement of infiltration directly over an entire watershed is impractical. A *soil mapping unit* is the smallest unit on a soil map that can be assigned a set of representative properties. The soil properties are stated in terms of layers. At a particular location on the map, because the properties of the soil vary with depth, some infiltration scheme is adopted for representing an essentially one-dimensional (vertical) process. The infiltration model representation may not include all layers used in the soil classification. Soil maps and the associated soil properties form a major source of data for estimating infiltration. A map originally compiled for agricultural purposes may be reclassified into infiltration parameter maps for hydrologic modeling. As such, the parameter map takes on a data structure characteristic of the original soil map. Figure 2.9 shows mapping units as polygons each with an identifier (map unit symbol, MUSYM). Associated soil properties within each polygon delienated are useful for modeling infiltration.

Estimating infiltration parameters from soil mapping units is introduced in Chap. 5. Other readily available digital maps may be used to derive parameters as discussed in the following section.

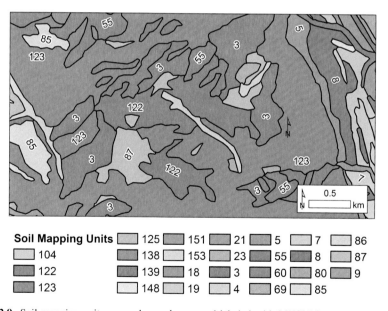

Fig. 2.9 Soil mapping units mapped as *polygons* and labeled with MUSYM

2.8   Land Use/Cover Classification

It is well known that land use, vegetative cover, and urbanization affect the runoff characteristics of the land surface. The combination of the land use and the cover, termed land use/cover, is sometimes available as geospatial data derived from aerial photography or satellite imagery. To be useful, this land use/cover must be reclassified into parameters that are representative of the hydrologic processes. Examples of reclassification from a land use/cover map into hydrologic parameters include hydraulic roughness, surface roughness heights affecting evapotranspiration and impervious areas that limit soil infiltration capacity. The data structure, raster or polygon, of the parent land use/cover map will carry into the model parameter map similar to how soil mapping units define the spatial variation of infiltration parameters.

Maps derived from remote sensing of vegetative cover affect the peak discharge and timing of the hydrograph in response to rainfall input. The hydraulic roughness parameter map takes on a data structure characteristic of the original land use/cover map. If derived from remotely sensed data, a raster data structure results rather than a polygonal structure. Figure 2.10 shows the polygonal data structure of a land use/cover map. This map is derived from the National Land Cover Database (NLCD) data updated in 2011 as described by Homer et al. (2012, 2015) and analysis of impervious cover presented by Xian et al. (2011). The unsupervised

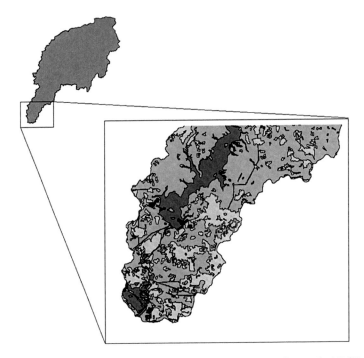

Fig. 2.10 Land use/cover for areas surrounding Lake Tenkiller according to the NLCD

classification scheme is derived from Landsat Thematic Mapper, producing the delineations shown in Fig. 2.10. The major categories are Open Water (11); Developed Low Density Residential (22); Developed Open Space (21); and Cultivated Crops (82). These land use classifications can be used to derive initial estimates of overland hydraulic roughness as discussed in Chap. 6.

2.9 Summary

Effective use of GIS data in distributed hydrologic modeling requires understanding of the type, structure, and scale of geospatial data used to represent watershed and runoff processes. GIS data often lack sufficient detail, space-time resolution, attributes, or differ in some fundamental way from how the model expects the character of the parameter or how the parameter is measured. Existing data sources are often surrogate measures that attempt to represent a particular category with direct or indirect relation to the parameter or physical characteristic. Generation of a surface from measurements taken at points, reclassification of generalized map categories into parameters and extraction of terrain attributes from digital elevation data are important operations in the preparation of a distributed hydrologic model using GIS. Having transformed the original map into usable parameter maps, the parameter takes on the characteristic data structure of the original map. Thus, the data structure inherent in the original GIS map has a lasting influence on the hydrologic process simulation using the derived parameters. In the following chapter, we turn to the generation of one data structure to another, surfaces from point data.

References

Band, L.E. 1986. Topographic partition of watershed with digital elevation models. *Water Resources Research* 22(1): 15–24.

Burrough, P.A. 1986. *Principles of Geographic Information Systems for Land Resources Assessment.* Monographs on Soil and Resources Survey, No. 12, 103–135. Oxford Science Publications.

FGDC. 1998. *Content Standard for Digital Geospatial Metadata.* rev. June 1998 Metadata Ad Hoc Working Group, Federal Geographic Data Committee, 1–78. Reston, Virginia.

Garbrecht, J., and L.W. Martz. 1997. The assignment of drainage direction over flat surfaces in raster digital elevation models. *Journal of Hydrologiques* 193: 204–213.

Homer, C.H., J.A. Fry, and C.A. Barnes. 2012. The national land cover database. *US Geological Survey Fact Sheet* 3020(4): 1–4.

Homer, C.G., Dewitz, J.A., Yang, L., Jin, S., Danielson, P., Xian, G., Coulston, J., Herold, N.D., Wickham, J.D., and Megown, K. 2015. Completion of the 2011. National land cover database for the conterminous United States-Representing a decade of land cover change information. *Photogrammetric Engineering and Remote Sensing* 81(5):345–354.

Jarvis, A., Reuter, H.I., Nelson, A. and Guevara, E. 2008. Hole-filled SRTM for the globe Version 4. *available from the CGIAR-CSI SRTM 90 m Database.* http://srtm.csi.cgiar.org.

Jenson, S.K., 1987. Methods and applications in Surface depression analysis. *Proceedings of Auto-Carto*: 8, 137–144. Baltimore, Maryland.

Jenson, S.K., and J.O. Dominique. 1988. Extracting Topographic Structure from Digital Elevation Data for Geographic Information System Analysis. *Photogrammetric Engineering and Remote Sensing* 54(11): 1593–1600.

Lee, J. 1991. Comparison of existing methods for building triangular irregular network models from grid digital elevation models. *International Journal Geographical Information Systems* 5(3): 267–285.

Moore, I.D., R.B. Grayson, and A.R. Ladson. 1991. Digital terrain modeling: A review of hydrological, geomorphological and biological applications. *Journal of Hydrologiques Process* 5: 3–30.

O'Callaghan, J.F., and D.M. Mark. 1984. The extraction of drainage networks from digital elevation data. *Computer Vision, Graphics and Image Processing* 28: 323–344.

Peucker, T.K., and D.H. Douglas. 1975. Detection of surface specific points by local parallel processing of discrete terrain elevation data. *Computer Vision, Graphics and Image Processing* 4(4): 375–387.

Rosenfeld, A., and A.C. Kak. 1982. *Digital picture Processing*. New York: Academic Press. No. 2.

Skidmore, A.K. 1990. Terrain position as mapped from a gridded digital elevation model. *International Journal of Geographical Information Systems* 4(1): 33–49.

Snyder, J.P. 1987. Map projections—A working manual. *U.S. Geological Survey Professional Paper* 1395: 383.

Turcotte, R.J.-P., A.N. Fortin, S. Rousseau, and J.-P.Villeneuve Massicotte. 2001. Determination of the drainage structure of a watershed using a digital elevation model and a digital river and lake network. *Journal of Hydrology* 240(3–4): 225–242.

Van Zyl, J.J. 2001. The shuttle radar topography mission (SRTM): a breakthrough in remote sensing of topography. *Acta Astronautica* 48(5): 559–565.

Xian, G., C. Homer, J. Dewitz, J. Fry, N. Hossain, and J. Wickham. 2011. Change of impervious surface area between 2001 and 2006 in the conterminous United States. *Photogrammetric Engineering and Remote Sensing* 77(8): 758–762.

Chapter 3
Surface Generation

Abstract The importance in hydrology of generating two-dimensional surfaces from data measured at points cannot be overstated. Transformation from point to raster data structures is often accomplished through interpolation of point data to form a two-dimensional surface. The difficulty with some interpolation methods is that the surface is organized around the point locations. The inverse distance weighting interpolation assigns a grid value based on the surrounding point values and distance away. If interpolated with the IDW method, it can appear that rain preferentially falls around the gauge, rather than having a more natural distribution geographically. It is important to understand how these surfaces are generated and the influence of the resulting surface on hydrologic modeling. The subject of this chapter is the methods and pitfalls of interpolation.

3.1 Introduction

Surface generation is accomplished by interpolating a grid-cell value from surrounding point values. Even if a surface is not required, point values may be required at locations other than at the measured point. This requires interpolation whether or not an entire grid is needed. Several surface generation utilities exist within general-purpose GIS packages such as ESRI ArcView, ArcGIS, or the public domain software called GRASS, Geographic Resources Analysis Support System (Neteler et al. 2012). The map shown above in Fig. 3.1 illustrates the difficulties associated with interpolating a hydrologic quantity such as annual rainfall depths measured at point locations.

Much work has been done in the area of spatial statistics, commonly referred to as *geostatistics* beginning with Krige (1951), Matheron (1965) who developed techniques for estimating quantities of precious metal in ore deposits. This type of statistics is distinguished by the fact that the data samples invariably are autocorrelated and are therefore statistically dependent at some range or length scale. Temperature, rainfall and topography, together with derived slope and aspect and infiltration rates or soil properties are measured at a point but are required in the raster data structure for modeling purposes.

B.E. Vieux, *Distributed Hydrologic Modeling Using GIS*,
Water Science and Technology Library 74, DOI 10.1007/978-94-024-0930-7_3

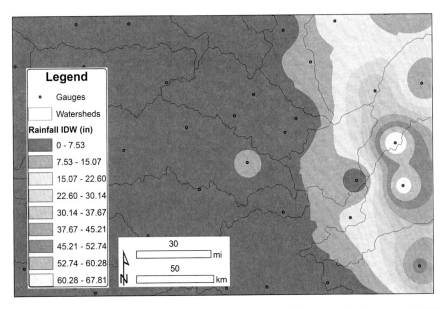

Fig. 3.1 Rainfall interpolated by the inverse distance weighting method shows rainfall erroneously distributed around gauge point locations

When generating a surface from irregularly spaced data points, we must choose whether it will be done by (1) interpolating directly on the irregularly spaced data, or by (2) regularizing the data into a grid and then applying some interpolation or surface generation algorithm. The first method is accomplished using points from LiDAR and then interpolating a TIN data structure, which has the advantage that *real* surface facets or break lines are preserved. TINs are interpolated surfaces and could logically be included here as a type of surface generation. The definition and characteristic data structure for TINs were discussed in Chap. 2 as a type of geospatial data structure. The second method has the advantage of producing a raster data structure directly with a smooth appearance. Smoothness, in the sense of differentiability, while desirable for many applications, may violate physical reality. In the following sections, we examine some of the more popular surface generation methods and application to interpolation of hydrologic quantities.

3.2   Interpolation

Interpolation from scattered data is a well-developed area of scientific literature with obvious applications in hydrology. A number of new mathematical techniques for generating surfaces have been developed. Although the surfaces generated by sophisticated mathematical approaches solve the interpolation problem, questions as to usefulness or practicality for various applications remain (Chaturvedi and

Piegl 1996). Surface generation is a more general case of interpolation in the sense that a value is needed in every grid cell of a raster array derived from point data. Among the many methods for generating a two-dimensional surface from point data, three will be addressed:

1. Inverse distance weighting
2. Thin splines
3. Kriging

The problem with all of these methods when applied to digital elevation models or geophysical fields such as rainfall, groundwater flow, wind, or soil properties is that some vital aspect of the physical characteristics becomes distorted or is simply wrong. The problems that arise when creating a surface from point data are:

- Gradients are artificially introduced by the algorithm
- Discontinuities and break lines, such as dikes or stream banks, may or may not be preserved
- Smoothing induced through regularized grids may flatten slopes locally and globally
- Interpolation across thematic zones such as soil type may violate physical reality.

Further difficulties arise when applying surfacing algorithms to point measurements of physical measurements such as soil properties. Interpolation or surface generation may extend across physical boundaries, resulting in physically unrealistic results. Because of this, it may be necessary to interrupt the smooth interpolated surface by break lines so that true linear features such as streams or edges of an embankment can be preserved.

The quantitative evaluation of the amount of spatial distribution of precipitation is required for a number of applications in hydrology and water resources management. Many techniques have been proposed for mapping rainfall patterns and for evaluating the mean areal rainfall over a watershed by making use of data measured at points such as the rain gauges shown above.

Methods for precipitation interpolation from ground-based point data range from techniques based on Thiessen polygons and simple trend surface analysis, inverse distance weighting, multiquadratic surface fitting and Delaunay triangulations to more sophisticated statistical methods such as the Barnes (1964) method that considers rain gauge network density. The weights applied to the surrounding gauge amounts are controlled by a parameter controlling the exponential decay of influence and distance from a gauge. When the exponential decay of influence is based on the mean gauge spacing, the interpolation method leads to the Barnes objective analysis (BOA) technique. Among statistical methods of interpolation, the geostatistical interpolation techniques such as kriging have been often applied to spatial analysis of precipitation, elevation, or other surface possessing some degree of autocorrelation.

Kriging requires the development of a statistical model describing the variance as a function of separation distance. Once the underlying statistical model is developed, the surface is generated by weighting neighboring elevations according to the separation distance. The underlying statistical model is perhaps the most critical and difficult task when kriging, since the choice of the model is somewhat arbitrary. The applicability or physical reality of the resulting model can hardly be assessed a priori.

Among any of the interpolation methods, smoothing across physical boundaries can result in a surface that is inconsistent with physical reality. Choosing appropriate methods for weighting neighboring values can range from functions based on simple distances, to application of more sophisticated polynomials called splines, or using a statistical model of autocorrelation. In the following sections, we will examine in more detail the application to hydrology of IDW, kriging, and splines for generating surfaces composed of elevations, precipitation, or some other hydrologic quantity.

3.2.1   Inverse Distance Weighted Interpolation

Inverse Distance Weighted (IDW) interpolation is widely used with geospatial data. The generic equation for inverse distance *weighted* interpolation is:

$$Z_{x,y} = \frac{\sum_{i=1}^{n} Z_i W_i}{\sum_{i=1}^{n} W_i} \qquad (3.1)$$

where $Z_{x,y}$ is the point value located at coordinates x,y to be estimated; Z_i represents the measured value (control point) at the ith sample point; and W_i is a weight that determines the relative importance of the individual control points in the interpolation procedure. How the weights are computed for the neighboring data values in the estimation of $Z_{x,y}$ is the principal difference among the methods considered.

The univariate IDW technique assumes that the surface between any two points is smooth but not differentiable; that is, this interpolation technique does not allow for an abrupt change in height. Bartier and Keller (1996) developed a multivariate IDW interpolation routine to support inclusion of additional independent variables. A control was also imposed on the abruptness of the surface change across thematic boundaries. Conventional univariate interpolation methods are unable to incorporate the addition of secondary thematic variables. Bartier and Keller (1996) found that the influence on surface shape did not allow for inclusion of interpretive knowledge. The multivariate IDW method of interpolation, on the other hand, supports the inclusion of additional variables and allows for control over their relative importance. The interpolated elevation, Z_{xy} is calculated by assigning weights to elevation values, Z_p found within a given neighborhood of the kernel:

$$Z_{x,y} = \frac{\sum_{p=1}^{r} Z_p d_p^{-n}}{\sum_{p=1}^{r} d_p^{-n}} \tag{3.2}$$

where Z_p is the elevation at point p in the point neighborhood r; d is the distance from the kernel to point p; and exponent n is the friction distance ranging from 1.0 to 6.0 with the most commonly used value of 2.0 (Clarke 1990). The negative sign of n implies that elevations closer to the interpolant are more important than those farther away. The size of r is an important consideration in preserving local features such as ridges, break lines or stream banks. Figure 3.2 shows the interpolation scheme where the grid cell elevation, $Z_{x,y}$ is interpolated from its nearest neighbors located at a distance, d_p, away from the grid cell of interest. The closer the neighboring value Z_p, the more weight it has in the interpolated elevation. The density of the point measurement and the variability of the point values affect the resulting surface. In the IDW method, the exponent, n and search radius or number of points, r in Eq. 3.2 are the only adjustable parameters that the user can select to fit the surface to the data. Visual inspection and validation are means for judging the value of the parameters that produce an optimal surface.

One artifact of the IDW method is the *tent pole effect* illustrated in Fig. 3.1 where rainfall is preferentially distributed around the point gauge locations. That is, a local minimum or maximum occurs at the measured point location. When applied to rain gauge accumulations, this gives the impression that it rained most intensely where there were measurements, which is clearly nonsensical. Figure 3.3 shows another more detailed surface interpolated from point rain gauge accumulations using the IDW method. The surface was interpolated using the inverse distance

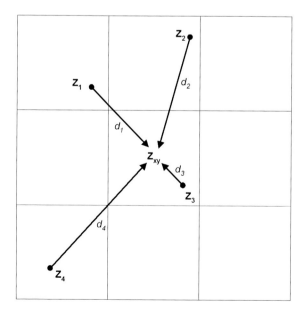

Fig. 3.2 Distance weighting scheme for interpolating grid cell values from nearest neighbors

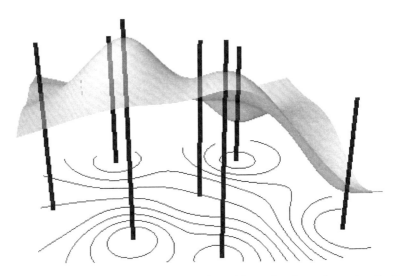

Fig. 3.3 Tent pole effect when generating a surface with too few data values using the IDW method

squared method, with $n = 2$ and using all data points. The contours from the rainfall surface are projected on to a plane, which indicates a physically unrealistic distribution of rainfall. The tent pole effect is problematic for most surfaces interpolated with the IDW method unless sufficient data points are obtained. Neither the IDW nor the other methods described below can insure physical correctness or fidelity in the generated surface. This surface was produced from point rainfall measured by seven rain gauges surround a 477 mi^2 (1,235 km^2) watershed for the purposes of estimating the basin area average rainfall for a storm event.

3.2.2   Kriging

One of the most important advantages of geostatistical methods such as kriging is the ability to not only interpolate a surface but to quantify the uncertainty of the surface. The basis for kriging is that statistical hypotheses are made to identify and evaluate the spatial dependence of the quantity measured at neighboring point locations. Kriging is an optimal spatial estimation method based on a model of spatially dependent variance. The model of spatial dependence is termed interchangeably a *variogram* or *semivariogram*. Surface interpolation by kriging relies on the choice of a variogram model. Inaccuracies due to noisy data or anisotropy cause errors in the interpolated surface that may not be obvious or be readily identified except through validation with data withheld from the surfacing algorithm. Todini and Ferraresi (1996) found that uncertainty in the parameters of the variogram could lead to three major inconsistencies:

1. Bias in the point estimation of the multidimensional variable
2. Different spatial distribution of the measure of uncertainty
3. Wrong choice in selecting the correct variogram model

For the kriging analysis, the spatial autocorrelation of data can be observed from a variogram structure. When the spatial variability is somewhat different in every direction, an anisotropy hypothesis is adopted. If no spatial dependence exists, kriging uses a simple average of surrounding neighbors for the interpolant. This implies that there is no dependence of the variance on separation distance.

The theory underlying kriging is well-covered elsewhere, especially in Journel and Huijbregts (1978) and subsequent literature. For those less familiar with the method, we examine the basic principles of kriging. In the theory of regionalized variables developed by Matheron (1965), there are two intrinsic hypotheses:

1. The mean is constant throughout the region.
2. Variance of differences is independent of position but depends on separation.

If the first hypothesis is not met, then a **trend** exists and should be removed. The second hypothesis means that once a variogram has been fitted to the data, it may be used to estimate the variance throughout the region. The semivariance, $\gamma(h)$, is defined over the couplets of $Z(x_i + h)$ and $Z(x_i)$ lagged successively by distance h:

$$\gamma(h) = \frac{1}{2N(h)} \sum [Z(x_i) - Z(x_i + h)]^2 \tag{3.3}$$

where $N(h)$ is the number of couplets at each successive lag. Figure 3.4 shows how grid data are offset by multiples of h, in this case, $2h$.

The variogram is calculated by sequentially lagging the data with itself by distance, h. The expected value of the difference squared is plotted with each lagged distance. The term *semi-* in 'semivariogram' refers to the fraction, ½ on the righthand side of Eq. 3.3.

Fig. 3.4 Lagging grid data by multiples of h producing semivariogram

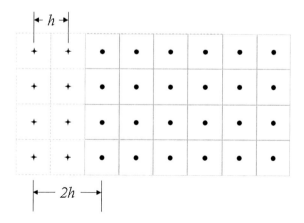

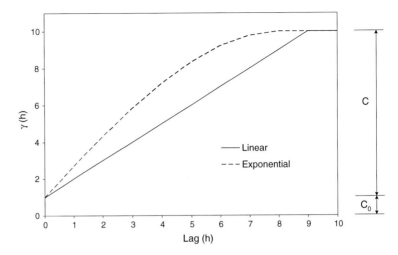

Fig. 3.5 Exponential and linear variogram models

Fitting a semivariogram model to the data is analogous to fitting a probability density function to sample data for estimating the frequency of a particular event. Once we have such a model, we can make predictions of spatial dependence based on separation distance. Only certain functions may be used to form a *model* of variance. These *allowable models*, (linear, spherical, or exponential among others) ensure that variance is positive. The linear and exponential semivariogram models in Fig. 3.5 establish the relationship between variance and spatial separation.

Beyond a distance equivalent to the range parameter, there is no dependence, resulting in the value of $\gamma(\infty) = s^2$, which is the variance in the classical statistical sense. At $h = 0$, i.e., no lag, the variation is termed the *nugget variance*. Because geostatistics had its start in mining and the estimation of gold ore, the term 'nugget' refers to the random occurrence of finding a *gold nugget*. In the more general geostatistical problem, this variance corresponds to a Gaussian process with no spatial dependence.

Linear model:

$$\gamma(h) = \begin{cases} C_0 + mh & 0 \leq h \leq a \\ C_0 + C & h \geq a \end{cases} \tag{3.4}$$

Exponential model:

$$\gamma(h) = \begin{cases} C_0 + C(1 - e^{-3h/a}) & 0 \leq h \leq a \\ C_0 + C & h \geq a \end{cases} \tag{3.5}$$

where C_0 is the nugget variance; C is the variance contained within the range, a. The variance C defines the variance as a function of the separation distance, h.

Once the statistical model, such as Eqs. 3.4 or 3.5, is fitted to the empirical semivariogram, the model is used to assign weights to data values neighboring the interpolant. Thus, kriging consists in statistical surface interpolation with known variance. The semivariogram provides the basis for weighting the neighboring data values. Unlike IDW, the semivariogram provides the kriging estimator with some basis for assigning weights to neighboring values besides distance. Weights for generating a Kriged surface $\hat{Z}(B)$ at location B is determined using weights taken from the semivariogram.

The simple kriging estimator within a restricted neighborhood of the interpolant, $\hat{Z}(B)$ at location B from n neighboring data points is:

$$\hat{Z}(B) = \sum_{i=1}^{n} \lambda_i Z_i \qquad (3.6)$$

where λ_i is the weight assigned to each of the elevations, Z_i. In kriging, the weights depend on the variogram model. Figure 3.6 shows how the interpolant is calculated given the distances and weights from the observed or measured values.

In practice, interpolated surfaces are more or less in error because either the underlying assumptions do not hold, or the variances are not known accurately, or both. An interpolated surface should be validated in any rigorous application. In geostatistical practice, the usual method of validation is known as *cross-validation*. It involves eliminating each observation in turn, estimating the value at that site from the remaining observations and comparing the two data sets. To be true cross-validation, the semivariogram should be recomputed and fitted each time an observation is removed. The main difficulty is that point data are usually so sparse

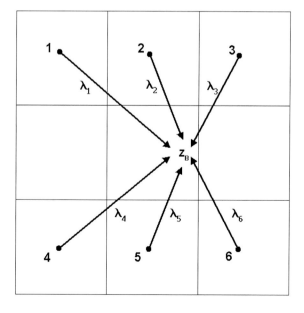

Fig. 3.6 Kriging interpolation based on weighted observations

that none can be omitted from the surface generation. Furthermore, refitting a semivariogram each time is cumbersome and time consuming. One difficulty with kriging and all surface generation techniques lies in having enough data to fit an acceptable model to and then validating the interpolated surface.

Validation can be performed by bootstrap methods, or by using a separate independent set of data for validation. Values are estimated at the sites in the second set and the predicted and measured values are compared. This is a true validation as described by Voltz and Webster (1990). The mean error, ME is given by:

$$\text{ME} = \frac{1}{m} \sum_{i-1}^{m} \{\hat{z}(x_i) - z(x_i)\} \tag{3.7}$$

where $z(x_i)$ is the measured value at x_i and $\hat{z}(x_i)$ is the predictand. For unbiased surface interpolation, the ME should be close to zero. MSE measures the precision of the prediction and, of course, it is preferred to be as small as possible. The mean square error, MSE, is defined as:

$$\text{MSE} = \frac{1}{m} \sum_{i-1}^{m} \{\hat{z}(x_i) - z(x_i)\}^2 \tag{3.8}$$

The mean square deviation ratio, R, is defined by:

$$R = \frac{1}{m} \frac{\sum_{i-1}^{m} \{\hat{z}(x_i) - z(x_i)\}^2}{\sigma_E^2} \tag{3.9}$$

where σ_E is the estimation variance for the predictand.

Interpolation of soil properties from point measurements can be accomplished by a variety of methods. Difficulties arise with most methods, kriging and splines included, especially where there are sharp discontinuities in soil classification. Accuracy of the interpolation method is affected by sampling density in relation to the spatial structure of the point measurements (Laslett et al. 1987; Laslett 1994; Voltz and Webster 1990; Zhu and Lin 2010). Several alternatives to ordinary kriging exist that can account for drift (nonstationarity) in the interpolant. When there are trends, they must be removed (detrended) or modeled within the spatial structure of the data. Methods such as *regression kriging* or *co-kriging* perform better than ordinary kriging (Hengl et al. 2007; Odeha et al. 1994, 1995)

An example of interpolating the rainfall shown above in Fig. 3.1 is repeated below in Fig. 3.7 using exponential kriging interpolation. With rain gauge locations, a large circular area is evident around an isolated gauge near the right center of the region. This rain gauge only partially reported rainfall during the period summed and thus appears anomalous.

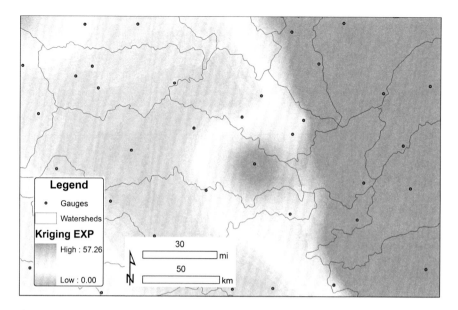

Fig. 3.7 Rainfall interpolated by the kriging method using an exponential model

Detrending data are important before attempting the identification of a semi-variogram model. Otherwise spurious correlations may become evident and may confound the interpolation. Recall that Matheron (1965) stated an intrinsic hypothesis that the mean is constant throughout the region, which implies that there are either no trends or they have been removed. Detrending can be applied to a point rain gauge and spatially distributed radar data. Figure 3.8 shows a 2 × 2 km gridded daily rainfall over Florida resulting from a hurricane. After removing trends within each 20 × 20 km block numbered 0–62, the data can be analyzed to determine an acceptable statistical model of the autocorrelation. The detrended data are shown in Fig. 3.9.

The range parameter found by robust fitting of exponential semivariograms to detrended daily radar data was found to be roughly 6 km (Vieux and Pathak 2007). Each storm event was modeled with a fitted semivariogram at daily time scales shown in Fig. 3.10 for a selected grid and the year 1995. The autocorrelation of daily rainfall was used to design a rain gauge network covering the water management district described in Vieux (2006), Vieux and Pathak (2007).

The exponential decrease in autocorrelation beyond the 6-km range parameter is evident from Fig. 3.10. The validity of the fitted variogram model should be verified in terms of the mean of standardized residual error (MRE) and variance of standardized residual error. We now turn to the spline method for generating a surface.

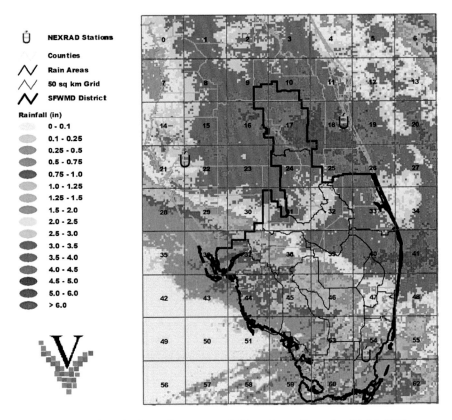

Fig. 3.8 Rainfall from Hurricane Frances recorded by radar on September 25–26, 2004

3.2.3 Spline

Splines are a class of functions useful for interpolating a point between measured values and for surface generation. *Cubic splines* are the most commonly used for interpolation. A one-dimensional cubic spline consists in a set of cubic curves, i.e., polynomial functions of degree three, joined smoothly end-to-end so that they and their first and second derivatives are continuous. They join at positions known as knots, which may be either at the data points, in which case the spline fits exactly, or elsewhere. In the latter case, the spline is fitted to minimize the sum of squares of deviations from the observed values and may be regarded as a smoothing spline. The number and positions of the knots affect the goodness of the fit and the shape of the interpolated curve, or surface in two dimensions. If the knots are widely spaced, local variations are filtered out. On the other hand, too many knots can cause the spline to fluctuate in a quite unwarranted fashion. To choose sensibly, the investigator must have some preconception of the kind of variation to expect, even

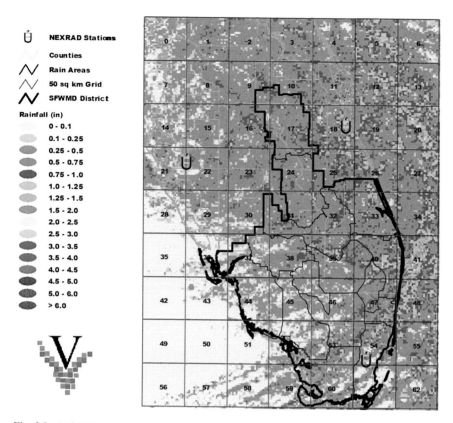

Fig. 3.9 As in Fig. 3.8 but detrended rainfall in each 20 × 20 km block

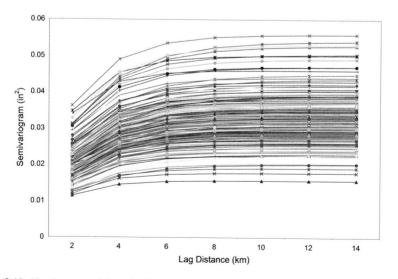

Fig. 3.10 Fitted exponential semivariograms from detrended rainfall (Vieux 2006)

though the technique demands fewer assumptions than other methods (Voltz and Webster 1990).

The *thin plate spline* (TPS) is defined by minimizing the roughness of the interpolated surface, subject to having a prescribed residual from the data. The TPS surface is derived by solving for the elevation, $z(x)$, with continuity in the first and second order derivatives:

$$z(x) = g(x) + h(x) \tag{3.10}$$

where $g(x)$ is a function of the trend of the surface, which depends on the degree of continuity of the derivatives; and $h(x)$ contains a tension term that adjusts the surface between the steel plate and membrane analogies. The mathematical details for solving this from a variational viewpoint are contained in Mitasova and Mitas (1993) and a comparison of splines with kriging, in Hutchinson and Gessler (1994).

The TPS method is a deterministic interpolation method using a mathematical analogy based on the bending of a thin steel plate. TPS is attractive because it is able to give reasonable looking surfaces without the difficulty of fitting a variogram. Moreover, comparative studies among different interpolators often find that deterministic techniques do as well as kriging. In this context, it is interesting to consider the ties of kriging to deterministic methods. In fact, some deterministic interpolators can be considered to have an implicit covariance structure embedded within them. Matheron (1981), for instance, elucidated the connections between the spline surface fitting technique and kriging and demonstrated that spline interpolation is identical in form to kriging with fixed covariance and degree of polynomial trend. Borga and Vizzaccaro (1997) found that a formal equivalence could be established between multiquadratic paraboloid surface fitting and kriging with a linear variogram model.

Spline methods are based on the assumption that the approximating function should pass as closely as possible to the data points and be as smooth as possible. These two requirements can be combined into a single condition of minimizing the sum of the deviations from the measured points and the smoothness of the function. Multidimensional interpolation is also a valuable tool for incorporating the influence of additional variables into interpolation, e.g., for interpolating precipitation with the influence of topography, or concentration of chemicals with the influence of the environment in which they are disturbed. The data smoothing aspects of TPS have also remained largely neglected in the geostatistical literature. The links between splines and kriging suggest that data smoothing methodology could be applied advantageously with kriging. Smoothing with splines attempts to recover the spatially coherent signal by removing noise or short-range effects.

Splines are not restricted to passing exactly through the data point. Data usually contain measurement error or uncertainty making this restriction unrealistic. The goal of surface interpolation is retrieval of the spatial trend in spite of the noise. Depending on the spline algorithm, various degrees of continuity can be enforced at each point. The resulting surface may be more like a membrane than a thin sheet of steel (*thin plate*), depending on the continuity in higher order derivatives.

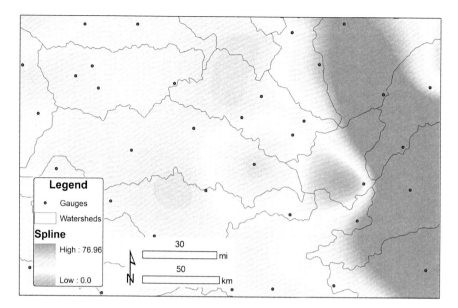

Fig. 3.11 Spline interpolation introduces variations not related to natural rainfall variation

Repeating the interpolation with splines, Fig. 3.11 shows the same annual rainfall depicted in Fig. 3.1 (IDW) and 3.7 (kriging) but with a tension spline produced in ArcGIS. Notice the same anomaly shows up but is less pronounced right of center.

Often, difficulties in making meaningful estimates from sparse data suggest the use of deterministic techniques such as splines. Unlike kriging, surfaces interpolated with splines do not have properties of *known variance*.

3.2.3.1 Generalizations of Splines

Thin plate splines are capable of several linear and nonlinear generalizations. Hutchinson and Gessler (1994) listed three direct generalizations of the basic thin plate spline formulations:

- **Partial thin plate spline** This version of splines incorporates linear submodels. The linear coefficients of the submodels are determined simultaneously with the solution of the spline. They may be solved using the same equation structure as for ordinary thin plate splines.
- **Correlated error structure** Using simple parametric models for the covariance structure of the measurement error, it can be shown that the spline and kriging estimation procedures give rise to solutions that are not consistent as data density increases. This is because highly correlated additional observations give essentially no further information. Nevertheless, spline and kriging analyses can result in a significant error reduction over the original data.

- **Other roughness penalties** A weakness of the basic thin plate spline formulation is that the spline has reduced orders of continuity at the data points. This effect may be countered by increasing the order of the derivative. Alternatively, the structure of the roughness penalty may be altered.

Spline surfaces that impose data smoothing yield interpolation errors similar to those achieved by kriging. Though the TPS method has not been used as widely in the geosciences as kriging, it can be used with advantages where the variogram is difficult to obtain.

There is no guidance as to how to adjust the parameters of the interpolation method, TPS or IDW, except by visual inspection or cross-validation. Experience with TPS indicates that it gives good results for various applications. However, overshoots often appear due to the stiffness associated with the thin plate analog. Using a tension parameter in the TPS enables the character of the interpolation surface to be tuned from thin plate to membrane. Mitasova and Mitas (1993) evaluated the effect of choosing both the order of derivative controlling surface smoothness and enforcement of continuity. The amount of data smoothing or tension is used to control the resulting surface appearance.

A class of interpolation functions developed by Mitasova and Mitas (1993) is *completely regularized splines* (CRS). The application of CRS was compared with other methods and was found to have a high degree of accuracy. TPS derived in a variational approach by Mitasova and Hofierka (1993) has been incorporated into the GRASS program call *s.surf.tps*. This program interpolates the values to grid cells from (x,y,z) point data (digitized contours, climatic stations, monitoring wells, etc.) as input. If irregularly spaced elevations are input, the output raster file is a raster of elevations. As an option, simultaneously with interpolation, topographic parameters slope, aspect, profile curvature (measured in the direction of steepest slope), tangential curvature (measured in the direction of a tangent to contour line) or mean curvature are computed.

The completely regularized spline has a tension parameter that tunes the character of the resulting surface from thin plate to membrane. Higher values of tension parameter reduce the overshoots that can appear in surfaces with a rapid change of gradient. For noisy data, it is possible to define a smoothing parameter. With the smoothing parameter set to zero (smooth = 0), the resulting surface passes exactly through the data points. Because the values of these parameters have little physical meaning, the surface must be judged by some independent means, such as cross-validation using the GCV method. Figure 3.12 shows an analysis wherein a tension parameter somewhere around 20 would minimize the cross-validation error. Note that smoothing does improve the surface for this particular data set.

The choice of surface generation method and corresponding parameters for a particular data set is somewhat subjective unless cross-validation is performed. It is not always clear which method or set of interpolation parameters yields the best results or most realistic surface. The resulting surface should be evaluated from a hydrologic perspective, which examines whether physical reality is violated, if

Fig. 3.12 Thin plate spline
cross-validation error
associated with smoothing
and tension parameters
(Mitasova and Mitas 1993)

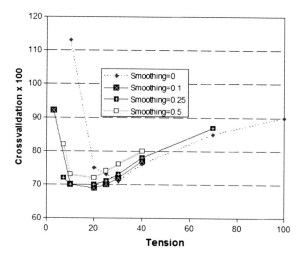

spurious local gradients are introduced, or if important hydrologic or geomorphic features are preserved.

3.2.3.2 Example of Surface Interpolation

Generating a topographic surface from surveyed data points for the purposes of distributed hydrologic modeling requires some choice of surface generation method. This section demonstrates the effects of smoothing and tension parameters on a surface generated using the thin plate spline algorithm, *s.surf.tps*. It is required to generate hydrographs using the digital terrain mapped interpolated from survey point measurements. This catchment was within the domain of a large-scale energy and hydrologic campaign, called HAPEX (Goutorbe et al. 1994). Characterizing rainfall runoff helps in understanding how and to what degree rainfall is transformed into runoff that collects in the valley bottoms called a kori. Small lakes (or *mares*) located in the kori are the major source of water in the region and are responsible for much of the recharge to a shallow groundwater table. Elevation data were obtained by surveying the land surface of the Wankama catchment, recording elevation and location in a relative coordinate system and then adjusting vertically to mean sea level and horizontally to the projected coordinate system, UTM.

The endoreic Wankama catchment (2 km^2) shown in Fig. 3.13 is representative of the discontinuous drainage network in this part of the Sahel. On the left (west), a large number of contours pass between the data points, resulting in a very rough surface. The calibration of the distributed parameter model, *r.water.fea*, for this basin relies on differential infiltration rates, higher rates for the ravines and lower values for crusted overland flow areas. The terrain was represented using a raster DEM derived from point elevation measurements. A topographic surface was then interpolated using the TPS algorithm (*s.surf.tps*). At first, a perplexing result

Tension 20

Tension 60

Tension 80

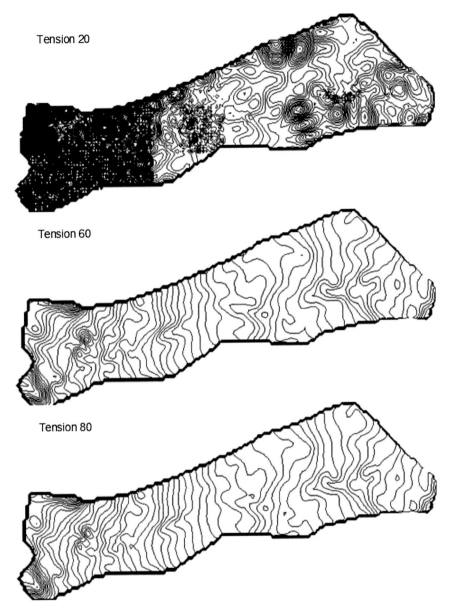

Fig. 3.13 Wankama catchment aerial extent draining from a plateau in the west to a sand-clogged valley bottom (kori and mare) in the east

presented itself, where very different surfaces result depending on the choice made for tension and smoothness parameters. The resulting DEM was used to derive the drainage network and slope for hydrologic modeling of rainfall runoff in this catchment. It is readily seen that the lower tension parameter equal to 40 (shown in gray) preserves some landscape features such as the ravine better than when the higher tension is equal to 80 (shown in black). A more detailed view of the plateau area to the western portion of the watershed reveals the differences in the surfaces generated with the 40 and 80 tension parameters. More ravine detail is seen in the contours produced using a tension parameter of 40 than with 80 (Séguis et al. 2002; Peugeot et al. 2003; Cappelaere et al. 2003; Desconnets et al. 1996).

Even though the surface passes through the data points, its physical reality is reduced. The black features that nearly cover the western portion of the watershed are closely spaced between surveyed elevation points. The tension parameter affects the smoothness of the resulting surface. As the tension parameter increases, the surface takes on a uniform appearance. However, this smoothing may remove from the surface actual landforms such as ravines. A tension parameter value should be chosen such that artifacts are removed while important landscape features are preserved.

3.3   Summary

Simulation of hydrologic processes requires extension of point measurements to a surface representing the spatial distribution of the parameter or input. With many surface generation techniques, unintended artifacts may be introduced. Gradients that are not physically significant may result along edges or where data are sparse. When using the IDW, the tent pole effect may be overcome by artificially making the number of points denser or by obtaining more measurements. The IDW has no fundamental means of control except to limit the neighborhood, in terms of either a radius or number of points used and the exponent (friction) used to interpolate the grid. Kriging, on the other hand, interpolates with weights assigned to neighboring values, depending on a spatially dependent model of variance. Control of the resulting surface is possible with the various spline algorithms.

TPS permits adjustment of smoothing and tension for more representative surfaces. Withholding measurements for validation of the surface permits some independent means of assessing whether the surface is representative of the physical feature. This is often difficult because all data are usually needed to create a representative surface. Visual inspection may be the only alternative to cross-validation in assessing whether a reasonable reproduction of the surface resulted from the surface generation algorithm. Surfaces or higher order gradients computed from the interpolated surface could potentially be erroneous due to anomalies introduced by incomplete, missing, or error prone point values.

A DEM derived by interpolation may be quite variable depending on seemingly arbitrary choices of smoothing and the interpolation method. Low smoothing and tension parameters generated unrealistic spatial gradients that may not be physically realistic. Too much smoothing and tension eliminate spatial variability, leading to a smooth surface that is visually pleasing but with many terrain features and variations that are either reduced or eliminated completely from the surface.

Whether these features are important to the ultimate goal of hydrologic modeling depends on the scale of the terrain features and the purpose of the modeling effort. If the purpose is to model erosion rates using the generated surface, local gradients due to the tent pole artifact may produce increased runoff velocities and higher erosion rates than are reasonable.

The ultimate goal is to generate a surface that reproduces the hydrologic character of the hydrologic quantity, rainfall, elevation, or similar measurement. This may entail creating several surfaces, deriving slope in the case of topography and then running a hydrologic model to observe the suitability of the surface for hydrologic simulation. In the next chapter, we will examine a means of assessing which resolution is sufficient for capturing the spatial detail contained in the data.

References

Barnes, S.L. 1964. A technique for maximizing details in a numerical weather map analysis. *Journal of Applied Meteorology* 3: 396–409.

Bartier, P.M., and P. Keller. 1996. Multivariate interpolation to incorporate thematic surface data using inverse distance weighting (IDW). *Computers & Geosciences* 22(7): 795–799.

Borga, M., and A. Vizzaccaro. 1997. On the interpolation of hydrologic variables: formal equivalence of multiquadratic surface fitting and Kriging. *Journal of Hydrology* 195: 160–171.

Cappelaere, B., B.E. Vieux, C. Peugeot, A. Maia, and L. Séguis. 2003. Hydrologic process simulation of a semi-arid, endoreic catchment in Sahelian West Niger, Africa: II. Model calibration and uncertainty characterization. *Journal of Hydrology* 279(1–4): 244–261.

Chaturvedi, A.K., and L.A. Piegl. 1996. Procedural method for terrain surface interpolation. *Computer & Graphics* 20(4): 541–566.

Clarke, K.C. 1990. *Analytical and computer cartography*, 290. New Jersey: Prentice-Hall, Englewood.

Desconnets, J.C., B.E. Vieux, B. Cappelaere, and F. Delclaux. 1996. A GIS for hydrological modelling in the semi-arid, HAPEX-Sahel experiment area of Niger, Africa. *Transactions in GIS* 1(2): 82–94.

Goutorbe, J.-P., T. Lebel, A. Tinga, P. Bessemoulin, J. Bouwer, A.J. Dolman, E.T. Wingman, J.H. C. Gash, M. Hoepffner, P. Kabat, Y.H. Kerr, B. Monteny, S.D. Prince, F. Saïd, P. Sellers, and J.S. Wallace. 1994. Hapex-Sahel: a large scale study of land-surface interactions in the semi-arid tropics. *Annales Geophysicae* 12: 53–64.

Hengl, T., G.B. Heuvelink, and D.G. Rossiter. 2007. About regression-kriging: from equations to case studies. *Computers & Geosciences* 33(10): 1301–1315.

Hutchinson, M.F., and P.E. Gessler. 1994. Splines—more than just a smooth interpolator. *Geoderma* 62: 45–67.

Journel, A.G., and C.J. Huijbregts. 1978. *Mining Geostatistics*. New York: Academic Press.

Krige, D.G. 1951. A statistical approach to some mine valuations and allied problems at the Witwatersrand, Master's thesis of the University of Witwatersrand.

Laslett, G.M., A.B. McBratney, P.H. Pahl, and M.F. Hutchinson. 1987. Comparison of several spatial prediction methods for soil pH. *Journal of Soil Science* 37: 617–639.

Laslett, G.M. 1994. Kriging and splines: An empirical comparison of their predictive performance in some applications. *Journal of American Statistical Association* 89: 391–409.

Matheron, G. 1965. *Les Variables Régionalisées et leur Estimation*. Paris: Masson.

Matheron, G. 1981. Splines and Kriging: their formal equivalence. In: *Down to Earth statistics: solutions looking for geological problems*. Syracuse University.

Mitasova, H., and J. Hofierka. 1993. Interpolation by regularized spline with tension: II. Application to terrain modeling and surface geometry analysis. *Mathematical Geology* 25(6): 657–669.

Mitasova, H., and L. Mitas. 1993. Interpolation by regularized spline with tension I. Theory and implementation. *Mathematical Geology* 25(6): 641–655.

Neteler, M., M.H. Bowman, M. Landa, and M. Metz. 2012. GRASS GIS: A multi-purpose open source GIS. *Environmental Modelling & Software*, 31: 124–130.

Odeha, I.O., A.B. McBratney, and D.J. Chittleborough. 1994. Spatial prediction of soil properties from landform attributes derived from a digital elevation model. *Geoderma* 63(3): 197–214.

Odeha, I.O., A.B. McBratney, and D.J. Chittleborough. 1995. Further results on prediction of soil properties from terrain attributes: heterotopic cokriging, and regression kriging. *Geoderma* 67(3): 215–226.

Peugeot, C., B. Cappelaere, B.E. Vieux, L. Séguis, and A. Maia. 2003. Hydrologic process simulation of a semi-arid, endoreic catchment in Sahelian West Niger, Africa: I. Model-aided data analysis and screening. *Journal of Hydrology* 279(1–4): 244–261.

Séguis, L., B. Cappelaere, C. Peugeot, and B.E. Vieux. 2002. Impact on Sahelian runoff of stochastic and elevation-induced spatial distributions of soil parameters. *Journal of Hydrological Processes* 16(2): 313–332.

Todini, E., and M. Ferraresi. 1996. Influence of parameter estimation uncertainty in Kriging. *Journal of Hydrology* 175: 555–566.

Vieux, 2006. *Rain Gauge Network Optimization Study*. West Palm Beach Florida: Report submitted to the South Florida Water Management District.

Vieux, B., and Pathak, C. 2007. Evaluation of Rain Gauge Network Density and NEXRAD Rainfall Accuracy. Proceedings of the American Society of Civil Engineers, World Environmental and Water Resources Congress, 1–12. doi:10.1061/40927(243)278.

Voltz, M., and R. Webster. 1990. A comparison of kriging, cubic splines and classification for predicting soil properties from sample information. *Journal of Soil Science* 41: 473–490.

Zhu, Q., and H.S. Lin. 2010. Comparing ordinary kriging and regression kriging for soil properties in contrasting landscapes. *Pedosphere* 20(5): 594–606.

Chapter 4
Spatial Variability Measuring Information Content

Abstract When building a fully distributed grid-based distributed hydrologic model from geospatial data, a fundamental question is: What resolution is adequate for capturing the spatial variability of a parameter or input? The model computational elements, whether finite element or difference, require parameter values that are representative of the grid cell. The composite for the model should then be representative of the spatial variation found in the watershed. When solving conservation of mass, momentum, or energy, a representative parameter value is required for that element size. The correspondence of the computational element size and the sampled resolution of the digital map affect how well the model will represent the spatial variation of parameters and modeled rainfall runoff. Within each grid cell, the conservation of mass and momentum is controlled by the parameters taken from special purpose maps whose values are hydraulic roughness, rainfall intensity, slope and infiltration rate for each computational element. The grid cell map supplies parameter values within the model grid. To take advantage of distributed parameter hydrologic models, the spatial distribution of inputs (e.g., rainfall) and parameters (e.g., hydraulic roughness) should be sampled at a sufficiently fine resolution to capture the spatial variability.

4.1 Introduction

Hydrologic simulations and model performance are affected by the seemingly arbitrary choice of a particular resolution. Additionally, in the processing of DEMs, smoothing or other filters are applied that may have deleterious effects on slope and other derived terrain attributes. This chapter deals with the resolution necessary or sufficient to capture spatial information that controls hydrologic response. Information content can be measured using a statistical technique developed in communications called informational entropy. Having a statistic that is predictive of the hydrologic impacts brought about by smoothing and resampling to another grid-cell resolution, especially DEMs, is quite useful. Figure 4.1 above shows the schematic for computing runoff in a 3×3 kernel. With this scheme it is possible to

© Springer Science+Business Media Dordrecht 2016
B.E. Vieux, *Distributed Hydrologic Modeling Using GIS*,
Water Science and Technology Library 74, DOI 10.1007/978-94-024-0930-7_4

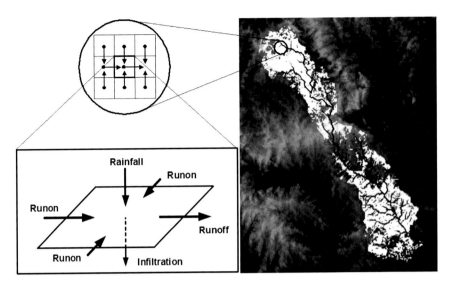

Fig. 4.1 Fully distributed grid-based runoff schematic for a 3 × 3 kernel in the Blue River Basin. The inset in the *upper right* shows spatially distributed runoff generated using a 270 m resolution grid

model an entire watershed using this elemental approach to building a watershed model.

Fully distributed grid-based hydrologic models require parameter values and precipitation assigned to each grid cell. The size of the grid cell determines how much spatial variability will be built into the model. Another way to view resolution is as a sampling frequency. Depending on the sampling frequency used and the spatial variability of the surface, the information may be under- or over-sampled. The goal is to capture the information content that is contained in a map of slope, soil infiltration parameters, surface hydraulic roughness, or rainfall. Each map and derivative parameter map is likely to have different resolutions that are necessary for capturing the spatial variability. It should be recognized that there is likely more variability than mapped. The measurement of spatial variability at a specific sampling frequency (resolution) is useful for deciding which resolution should be used for each map. Practical considerations involve finding the smallest grid-cell size and then using that for the other maps even if this involves oversampling less variable parameter maps.

The resolution that is sufficient to capture the variability of a parameter depends on the spatial variability of the values. If the parameter is actually constant, then any resolution is adequate for sampling the surface. However, the more likely case is that the parameter or input is spatially variable, requiring a decision as to the most efficient choice of resolution. Opting for the smallest resolution possible wastes storage space and computational efficiency. Depending on the extent or size of the river basin, the resolution that adequately represents the spatial variability important to the hydrologic process may range from tens or hundreds of meters for hillslopes to hundreds or thousands of meters for river basins. For very large basins, even larger resolutions may be appropriate. The real question is whether we can use larger resolutions to minimize the use of computational resources and still represent the spatial variability controlling the response.

Spatial data are widely recognized as having a high degree of autocorrelation. That is to say, adjacent cells tend to have values similar in magnitude. The *correlation distance* is the length scale that separates whether sampled values appear to be correlated or independent. This same concept was dealt with in Chap. 3, where the range of spatial dependence in the kriging method determined the weights for neighboring data values. Sampling at resolutions greater than this length will produce variates that are independent, showing no high degree of autocorrelation. Sampling at resolutions smaller than this length will produce variates with a high degree of autocorrelation. Hengl et al. (2007) presented a comprehensive set of decision criteria for grid resolution selection including a spatial autocorrelation structure, information content and spatial dependence.

Derivation of parameter maps from thematic classifications, e.g., soil or land use, will appear to be constant within the polygon delineating the class. More variation undoubtedly exists than is mapped. The within-class variation may or may not be important to the simulation of the hydrologic process. For relatively large grid cells more than 500 m, subgrid parameterization may be required for any runoff to be generated from a land surface model. Incorporating a distribution of model parameters rather than the mean has been found to improve WRF-HYDRO runoff simulation as suggested by Mendoza et al. (2015). Rather than introducing spatially uncorrelated parameter distributions, spatial structure can be incorporated into the subgrid parameter. For example, Wang et al. (2015) described a geostatistical approach for area-to-point regression kriging (ATPRK). This downscaling of MODIS satellite data at 500 m resolution provided improved monitoring of deforestation. Subgrid parameterization can also be improved by incorporating a probability distribution for the parameter. Downscaling can also leverage the introduction of spatial autocorrelation, rather than assuming a constant parameter value for the grid that is representative of the thematic polygon. Cell size selection should consider the information content captured by a given resolution.

Many of the perceived problems with physics-based distributed models are difficulties associated with parameterization, validation and representation of within-grid processes. However, a strength of physics-based models is the ability to incorporate the spatial variability of parameters controlling the hydrologic process. However, as Beven (1985) pointed out, the incorporation of spatial variability in physics-based models is possible only "…if the nature of that variability were known." The 'nature of variability' is a qualitative term describing the kind of variability, in contrast to quantitative terms such as 'variance' or 'correlated length' that describe the amount of variability. Specifically, the nature of spatial variability may be represented as deterministic and/or stochastic (Smith and Hebbert 1979; Philip 1980; and Rao and Wagenet 1985). The total variability of a given parameter is a composite of the deterministic and stochastic components. Within-class variation is the stochastic component, whereas between-class variation is deterministic. An important special case is when homogeneity, or no variability, is assumed. In reality, spatial variability is rarely entirely deterministic or stochastic. Within any deterministic trend or distribution of parameter values, there is invariably some degree of uncertainty or stochastic component. Similarly, stochastic variation may be nonstationary; containing systematic trends, or is composed of nested, deterministic variability.

A number of studies have demonstrated that stochastic variability can have a large impact on hydrologic response. In particular, those studies showed that processes such as runoff and infiltration, which result from a hypothetical random distribution of parameters, are often not well represented by homogeneous effective parameters. This has led to stochastic modeling approaches and uncertainty prediction. In addition to work on stochastic variability, Smith and Hebbert (1979) demonstrated that deterministic variability may have important impacts on surface runoff. Another view is to recognize some limiting size (Wood et al. 1990) of a subbasin or computational element called a representative elementary area (REA) to establish a 'fundamental building block for catchment modeling.' Smith and Hebbert (1979) noted, however, that the deterministic length scale depends on the scale of interest, the processes involved and the local ecosystem characteristics. Introducing new sources of deterministic variability related to topography, soils, climate, or land use/cover might be accomplished by mapping at a finer resolution and scale.

Seyfried and Wilcox (1995) examined how the nature of spatial variability affects the hydrologic response over a range of scales, using five field studies as examples. The nature of variability was characterized as either stochastic, when random, or deterministic, when due to known nonrandom sources. In each example, there was a deterministic length scale, over which the hydrologic response was strongly dependent upon the specific, location-dependent ecosystem properties. Smaller scale variability was represented as either stochastic or homogenous with nonspatial data. In addition, changes in scale or location sometimes resulted in the introduction of larger scale sources of variability that subsumed smaller-scale sources. Thus, recognition of the nature and sources of variability was seen to reduce data requirements by focusing on important sources of variability and using nonspatial data to characterize variability at scales smaller than the deterministic length scale.

The existence of a deterministic length scale implies upper and lower scale bounds. At present, this is a major limitation to incorporating scale-dependent deterministic variability into physics- based models. Ideally, the deterministic length scale would be established by experimentation over a range of scales. This, however, will probably prove to be impractical in many cases. In general, the lower boundary will be easier to estimate. Although it is difficult to define the nature of spatial variability with respect to sources of variability, it is important to recognize that these definitions are effectively made during model design and application. The determination to include certain kinds of elevation data or use a certain grid spacing, for example, implies specific treatments, whether intended or not, of spatial variability. These aspects of model design are more likely to be driven by computation time and data availability than by spatial variability. Even when driven by spatial variability, these considerations are not usually stated up front as part of the model. This makes interpreting model results and transferring the model to other locations difficult.

Recognition of the nature of variability has other important implications for physics-based modeling. Although physics-based models are generally comprehensive in their consideration of the basic hydrologic processes, they may fail to account for sources of variability that largely control these processes. This failure may be due to grid sizes that are too large or to lack of appreciation of the impact of different variability sources. Dunne et al. (1991), for example, noted the lack of appreciation of shrub influences on snow drifting in alpine hydrologic modeling. Tabler et al. (1990) made similar observations concerning snow drifting. With regard to soil freezing, Leavesly (1989) noted that frozen soil was identified as a major problem area but none of the 11 models that were included in the study had the capability to simulate frozen soils. The important advantages of physics-based models over other approaches cannot be expected to accrue unless the effects of critical sources of variability are taken into account. The degree to which a model takes into account critical spatial variability has important effects on model data requirements, grid scale and measurement scales.

Whether a soil is frozen at a particular location and time or the location of shrubs in a landscape verges on unknowability, exemplifies the difficulty of distributed modeling. While the following considerations enumerated by Leavesley (1989) pertain to specific hydrologic applications, they should be taken into account when developing or applying a distributed hydrologic model.

- **Data requirements**: The effects of scale on spatial variability suggest why virtually unlimited data (Freeze and Harlan 1969) and computer demands may not have to be met to simulate hydrologic processes on a physical basis. Increasing the scale of interest introduces additional sources of variability whose effects on hydrologic response may subsume those from smaller scales. In this context, greater data quality may not mean greater accuracy but simply mean more information.
- **Grid scale**: Another important consideration is the model grid size. Physics-based models generally account for spatial heterogeneity with inputs at nodes of a grid system. Grid point parameters are generally assumed to represent a homogeneous grid, which results in empirically effective parameters. This approach is necessary for computational reasons and allows for the incorporation of a variety of data inputs via geographical information systems. Grid size should be determined within the context of the nature of the variability. The grid must be small enough to describe important deterministic variability.
- **Measurement scale**: Many of the problems associated with spatial variability and scale have been related to scale of measurement. It has been suggested that the development of large-scale measurement techniques, by remote sensing, for

example, will provide model input and verification, which will largely eliminate problems related to spatial variability (Bathurst and O'Connell, 1992).

If a particular data set is undersampled, important variation in space will be missed, causing an error in the model results. Oversampling at too fine a resolution wastes computer storage and causes the model to run more slowly. The ideal is to find the resolution that adequately samples the data for the purpose of the simulation yet is not so fine that computational inefficiency occurs.

We will assume that the GIS system used to perform the hydrologic analysis of input data has the capability for maintaining each map with its own resolution. The topography and the derived parameters such as slope may be at a relatively small resolution dictated by the available DEM, say on the order of 30 m, whereas precipitation estimates derived from radar or satellite may be at 4 km or larger. Identifying macroscale parameters that are 'effective' for the resolution was investigated by Wood (1995). Sampling a variable surface successively means coarser resolution can be performed to assess the sufficient resolution for capturing variation. Figure 4.2 shows the IDW rainfall surface (see Chap. 3) along with sample points shown at 2,000 and 4,000 m intervals.

Because the rainfall represents annual totals the increased resolution at 2,000 m or even 4,000 m is sufficient to capture the spatial variability. As shown in the histogram in Fig. 4.3 when the resolution increases to 8 and 16 km, several

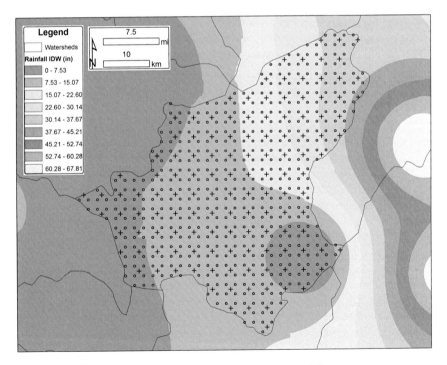

Fig. 4.2 Sampling locations on an interpolated map of annual rainfall

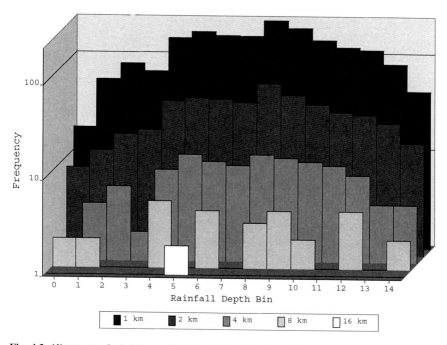

Fig. 4.3 Histogram of rainfall sampled at resolutions from 1 to 16 m from the interpolated surface

categories are not represented by the sampled resolution, e.g., at 8 and 16 km. As seen in Fig. 4.3 when the rainfall surface is sampled from 500 to 4,000 m resolution the frequency of cells (histogram bins) is seen to shift slightly and become less clear as to what the mode of the rainfall is. Whether the precipitation is adequately sampled for the size of the watershed depends on relative scales and the variability of the surface.

Determining the resolution that is sufficient for representing the variation of the mapped quantity can be done empirically by testing how the central tendency shifts as resolution is increased to larger sizes. A statistical measure of the surface variation that can be tested across a range of resolutions could be as simple as descriptive statistics or the information content approach described next.

4.2   Information Content

The goal of spatial information content measurement is to have some measure that indicates whether the resolution of a data set oversamples, undersamples or is simply adequate to capture the spatial variability. We wish to be able to tell whether the resolution is sufficient to capture the information contained in a map of slope, soil infiltration parameters, surface hydraulic roughness, or rainfall. Avoiding excessively small resolutions that do not *add information* can be accomplished by measuring information content at successively larger or smaller resolutions. Identifying the resolution that is sufficiently small to capture important details is answered in part by testing the quantity of information contained in a data set as a function of resolution. Once the information content ceases to increase as we resample the original data at a finer resolution, we can say that we have captured the information content of the mapped parameter and finer resolution is not merited or supported. Borrowing from communication theory, information content is a statistic that may be computed at various resolutions (Vieux 1993). Such a measure can test which resolution is adequate in capturing the spatial variability without the need for fitting a probability distribution function to often complex distributions typical of spatial data.

Smoothing and resampling reduce the spatial variability of elevation and the derived slope data. Smoothing and resampling to coarser resolution are essentially data filters that reduce information content. The application of information theory to determining bandwidth needed in a communication channel was first introduced in the landmark theory by Shannon and Weaver (1964). Application of information content to hydrologic modeling was found to be useful by Vieux (1993); Vieux and Farajalla (1994); Farajalla and Vieux (1995) and Brasington and Richards (1998) among others. Information content helps in understanding the effects of data filters or resolution. Informational content or entropy, *I*, in this context is defined as:

$$I = -\sum_{i=1}^{\beta} P_i \log(P_i) \tag{4.1}$$

where β is the number of bins or discrete intervals of the variate and P_i is the probability of the variate occurring within the discrete interval. A negative sign in front of the summation is by convention such that increasing information content results in positively increasing informational entropy. A base 10 logarithm is used in this application, yielding units of Hartleys. Information content in communication theory commonly uses base 2 when operating on bits of information yielding *binits* of information. For a complete description of informational entropy in a communication theory context, the reader is directed to Papoulis (1984).

Informational entropy becomes a measure of spatial variability when applied to topographic surfaces defined by a raster DEM. As the variance increases, so does informational entropy. Conversely, as variance decreases, so does informational entropy. In the limit, if the topographic surface is a plane with a constant elevation, the probability is 1.0, resulting in zero informational entropy, zero uncertainty and zero information content. Maximum informational entropy occurs when all classes or histogram bins are equally probable. With all bins in the histogram filled, every elevation is equally probable in the DEM. Measuring informational entropy at increasing resolutions provides an estimate of the rate of information loss due to resampling to larger cell sizes.

4.3 Fractal Interpretation

The rate of information loss with respect to cell size can be put into terms of a noninteger or fractal scaling law. As the sampling interval increases with increasing cell size, the information loss is greater for surfaces with a higher fractal dimension. In turn, the information loss propagates errors in the hydrologic model output. Thus, information content is an indication of how much spatial variability is lost due to filtering by either smoothing or aggregation. The rate of information loss would vary in any natural topography. High variance at a local scale demands smaller cell sizes to capture the information content. When variance is low (a surface with constant elevation), any number of cells is sufficient to capture the information content including just one cell. For example, a plane surface will have a fractal dimension equal to the Euclidian dimension of 2. A completely flat or uniform slope will not lose information when smoothed or resampled because there is no spatial variability to be lost. On the other hand, topographic surfaces possessing high variability with a fractal dimension greater than 2.0 will experience a higher degree of information loss as smoothing and resampling progress.

The information dimension is the fractal dimension computed by determining the information content as measured by informational entropy at different scales. Several definitions of dimension, including the information dimension, for dynamic systems exhibiting chaotic attractors, were presented by Farmer et al. (1983). The fractal dimension is a measure common to many fields of application. Information content measured at different scales is not a self-similar fractal, because a self-similar fractal exhibits identical scaling in each dimension. In terms of a box dimension, a rectangular coordinate system may be scaled down to smaller grid cells. Following the arguments of Mandelbrot (1988), a *self-affine* function (such as informational entropy) follows the nonuniform scaling law where the number of grid cells is scaled by ε^H. Writing the scaling law in terms of a proportionality, the number of grid cells, $N(\varepsilon)$, of side ε is:

$$N(\varepsilon) \propto \varepsilon^H \qquad (4.2)$$

where ε is the grid dimension and H is the Hurst coefficient. Note that the fractal dimension, d_I, must exceed the Euclidian dimension and that $0 \le H \le 1$.

Therefore:

$$d_I = E + 1 - H \qquad (4.3)$$

The variable, H, has special significance (Saupe, 1988): at $H = 0.5$ the surface is analogous to ordinary Brownian motion, when $H < 0.5$ there is a negative correlation between the scaled functions and for $H > 0.5$, there is a positive correlation between the scaled functions.

The information dimension described by Farmer et al. (1983) is based on the probability of the variate within each cube in three dimensions or within a grid cell

Table 4.1 Information content of a plane surface with uniform slope

E	β	I(ε)
1/625	625	2.7958
1/125	125	2.0969
1/25	25	1.3979
1/5	5	0.6990

in two dimensions covering the set of data. As a test case, we examine the dimension of a plane. The concept is extended to the more general topographic surface of a watershed basin in subsequent sections. The plane surface is subdivided into cells of ε = 1/5th, 1/125th and 1/625th. The plane surface is of uniform slope such that each elevation has an equal probability of $I(\varepsilon) = \log(N(\varepsilon))$. If a plane surface of uniform slope in the direction of one side of the cells is divided into a five by five set of cells, it will have five unique rows of equally probable elevations. There are five categories of elevation in the DEM and five bins. The number of bins is β. Each bin will have a probability of 1/5 and an informational entropy of 0.69897, which is simply $\log(B)$. If the plane is divided into successively smaller grid cells, the rate at which the informational entropy changes should be equal to one for a plane surface. Table 4.1 presents the calculations for the plane of uniform slope and one cell wide. Under the condition of equal probability among classes and exactly one occurrence per class or interval, the number of bins, B equals the number of cells N for a unit width plane.

The Hurst coefficient, H, is the proportionality factor relating the number of grid cells, $N(\varepsilon)$, to the size of the cells (ε). We expect that informational entropy scales with the size of the cell of side ε covering the set according to Eq. 4.1. Taking logarithms of both sides of Eq. 4.1, replacing the proportionality with a difference relation and recognizing that for equal probabilities, $\log(N(\varepsilon)) = I(\varepsilon)$, we find that H is the proportionality constant that relates the rate at which informational entropy changes with grid-cell size and is related to the fractal dimension, d_I by:

$$H = \frac{\Delta I(\varepsilon)}{\Delta \log(1/\varepsilon)} \tag{4.4}$$

where $I(\varepsilon)$ is the information content computed for grid cells of side ε. Applying Eq. (4.4) to the values in the last col. in Table 4.2, we find that:

$$H = (2.796 - 2.097)/(\log(625) - \log(125)) = 1 \tag{4.5}$$

Thus, for a planar two-dimensional surface with no variability, E = 2 and H = 1 and by Eq. (4.3), d_I = 2, as would be expected for a plane surface. Furthermore, H = 1.0, which is greater than 0.5, which indicates that elevations are positively correlated. That is, similar elevations are found close together, a feature that is typical of real terrain.

The information dimension is a measure of the rate at which information changes with resolution or by some other filter such as smoothing. Table 4.1 demonstrates that the information dimension measures the rate at which information changes for a plane surface; $H = 1$ and $d_I = E$. It also shows that any resolution is sufficient to capture the spatial variability. We have demonstrated how resolution affects the information content sampled by a range of resolutions for an idealized surface. Next, we turn our attention from a theoretical plane surface with uniform slope to the effect of resolution on an actual DEM that is used to derive slope.

4.4 DEM Resolution Effects

Resampling a DEM at larger resolutions means that we take the original resolution and increase the resolution by selecting the cell closest to the center of the new larger cell size. If we increase by odd multiples of the original resolution, say 3×3 or 5×5 windows, the center cell of the window is used as the elevation for the new larger cell. As resampling progresses, the topographic data become a data set of larger and larger cells. Resampling is a means of investigating the effects of using larger cell sizes in hydrologic modeling. The appropriate cell size or resolution is the largest one that still preserves the essential characteristics of the digital data and its spatial characteristics or distribution. To gain an appreciation for how information content is related to spatial variability, we start with a plane surface. Informational entropy applied to spatially variable parameters/input maps is useful in assessing the resolution that captures the information content. Special considerations for different types of data are addressed in the following applications.

The choice of resolution has an important impact on the hydrologic simulation. Choosing a coarse resolution DEM, deriving slope and using this in a surface runoff model has two principal effects. One is to shorten the drainage length, because many of the natural meanders or *crookedness* of the drainage network are short-circuited by connecting raster grid cells together by way of the principal slope. The effect of grid-cell resolution on the drainage network is addressed in Chap. 7. The other effect is a *flattening* of the slope due to a sampling of the hills and valleys at too coarse a resolution. We can think of it as cutting off the hilltops and filling in the valleys. These effects together may have compensating effects on the resulting hydrograph response. A *shortened* drainage length decreases the time taken by runoff from the point of generation to the outlet. A *flattened* slope will increase the time.

Brasington and Richards (1998) examined the effects of cell resolution on TOPMODEL and found that information content predicted a break in the relation between model response and resolution. Sensitivity analyses revealed that model predictions were consequently grid-size dependent, although this effect could be modulated by recalibrating the saturated hydraulic conductivity parameter of the model as the grid size changed. A significant change in the model response to scale was identified for grid sizes between 100 and 200 m. This change in grid size was also marked by rapid deterioration of the topographic information contained in the DEM, measured in terms of the information content. The authors suggested that this break in the scaling relationship corresponds to typical hillslope lengths in the dissected terrain and that this scale marks a fundamental natural threshold for DEM-based application.

Quinn et al. (1991) presented the application of TOPMODEL, which models subsurface flow at the hillslope scale. Model sensitivity to flow path direction derived from a DEM was investigated. The application by Quinn et al. (1991) used a 50 m grid-cell resolution, which is the default value of the United Kingdom data base. Resampling at larger grid-cell resolutions was found to have significant effects on soil moisture modeling due to aggregation.

The drainage network extracted from a DEM is affected by DEM resolution. Tarboton et al. (1991) investigated stream network extraction from DEMs at various scale resolutions. They found the drainage network density and configuration to be highly dependent on smoothing of elevations during the pit removal stage of network extraction. In fact, if smoothing was not applied to the DEM prior to extraction of the stream channel network, the result did not resemble a network. While smoothing may be expedient, it can have deleterious effects, viz., undesirable variations in a delineated watershed area.

Relating the quantitative effects of data filters to hydrologic simulation was reported by Vieux (1993). The effects on hydrologic modeling produced are measured using information content affected by two types of filters: smoothing and cell aggregation. As mentioned above, cell size selection is important in capturing the spatial variability of the DEM. Smoothing is often necessary before automatic delineation of the watershed and stream network to reduce the number of spurious high or low points, referred to as *pits* and *spikes*. Both smoothing and resampling to larger resolutions have the effect of flattening the slope derived from such a filtered DEM. A nearly linear relationship was found between a delayed hydrograph response and the relative informational entropy loss due to both smoothing and aggregation. The error introduced into the simulation was not constant for all rainfall intensities. Higher intensity events revealed less change in response to filtering than did less intense events. Because the more intense events achieved equilibrium more quickly, the changes in spatial variability had less effect. Though rare in nature, an equilibrium hydrograph means that input equals output. When input (rainfall excess) equals output (discharge at the outlet), effects of spatial variability in the output are no longer present. To demonstrate the influence of resolution effects on elevation and slope, a series of coarser DEMs were used to derive slope. The decrease in the average slope is seen in Fig. 4.4.

Analysis of the DEM resampled from 30 to 60 m shows that the derived slope maps decrease slightly from 3.34 to 3.10 %. This may be an acceptable degree of flattening, given that computer storage decreases to $30^2/60^2$ or ¼ of that required to store the 30 m DEM. At 960 m resolution, there is a considerable reduction in slope with essentially one slope class around 2.5 % and therefore, zero information entropy indicating a loss of information content. Little or no loss of information content was found for elevation during the resampling. This means that at these resolutions, equivalent distributions of elevations occur but the first derivation, slope changes markedly. Besides reduced storage, subsequent computational effort

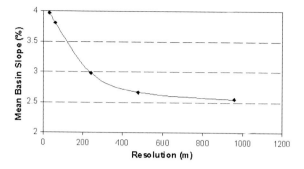

Fig. 4.4 Slope flattening due to increased DEM resolution

will be more efficient at larger resolution. A physical interpretation is that the hillslopes are not adequately sampled, resulting in a cutting of the hilltops and filling of the valleys.

As slope is derived from coarser resolution DEMs, the steeper slopes decrease in areal extent and are reflected in the decrease in mean slope. Figure 4.5 shows four maps produced from resampled DEMs. The darker areas are the steeper slopes (>10 %). Flatter slopes in the class 0–1 % also increase, as is evident particularly in

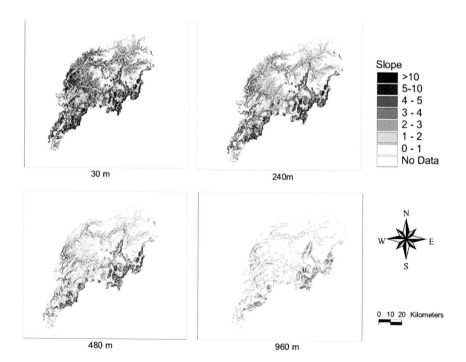

Fig. 4.5 Slope maps derived from 30, 240, 480 and 960 m DEMs

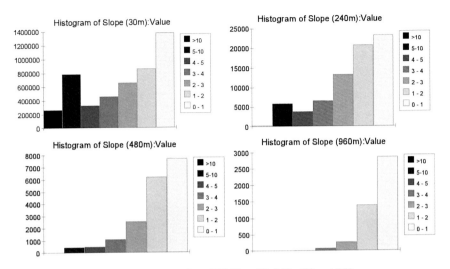

Fig. 4.6 Histograms of slopes derived from DEMS at 30, 240, 480 and 960 m

the upper reaches of the basin. Information content is reduced primarily because the bins in the histogram are not filled.

This same effect of resolution-induced flattening can be seen in Fig. 4.6, which reveals how steeper slope classes have fewer counts than flatter classes. As steeper slopes drop out of the distribution, they do not contribute to the information content statistic. At the same time, the average watershed slope becomes flatter with coarser resolution.

The information content loss decreases on a relative basis compared to the finest resolution, as well as the mean slope. Figure 4.7 shows a nearly linear relationship ($r^2 = 0.96$) between the relative loss in mean slope and information content.

Relative loss means the amount lost in relation to the information entropy or slope at 30 m resolution. The decrease in mean slope is about two times the loss in information content. By resampling, if the information content decreases by 100 %, then a 50 % decrease in mean slope should be expected. The rate of decline for

Fig. 4.7 Relative loss in slope in relation to loss in information content

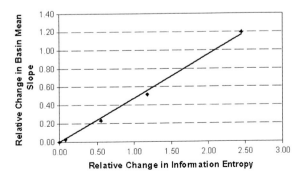

other regions will depend on the initial starting resolution and the length scales of the topography in relation to resolution.

Information content is useful for characterizing the spatial variability of parameters affecting distributed hydrologic modeling of surface and subsurface runoff, where hydraulic conductivity is a key parameter in Darcy flow governing the velocity of flow and its arrival time at an outlet. Niedda (2004) found a relationship between information content loss and hydraulic conductivity as the DEM grid cell resolution was resampled to coarser sizes. As a result of increasing grid size, soil hydraulic conductivity had to be increased, to compensate for the resulting retardation of the runoff hydrograph. This correlation and its relation to information content loss were exploited to upscale the hydraulic conductivity parameter with increasingly coarser grid cell resolution. The relationship between K at a selected grid cell size and K^* at the point scale (1 m resolution) estimated from soil texture is expressed in terms of information content loss, ΔI, as:

$$K = K_S^* e^{\Delta I} \qquad (4.6)$$

An apparently strong correlation of hydraulic conductivity with terrain curvature helps to establish a predictable relationship between the optimized parameter and grid cell resolution. Larger K values are needed with coarser grid cell resolution to compensate for the attenuated hydrograph response that may be related to grid-cell dependence rather than uncertainty or knowledge about the soil properties at the catchment scale. For the basins tested, information content loss was a useful index of parameter rescaling. Jana and Mohanty (2012) observed similar effects on hydraulic conductivity, where the optimized soil hydraulic parameter, K values, must increase as grid cells are upscaled to coarser sizes. Recognizing the scale dependence of parameter values is an important step in distributed modeling since grid cell resolutions are often chosen that are larger than the scale at which parameter values were measured or derived.

Wise (2012) used information content, or entropy, to judge the quality of a DEM by testing the representation of elevation classes produced by various surface interpolation methods such as inverse distance weighting or splines. In Wise (2011), one of the more sensitive characteristics derived from a DEM was found to be slope and curvature (first and second derivatives). The slope (gradient) tended to be underestimated when low-resolution DEMs were used. Additional discussion of the effects of DEM resolution, smoothing and interpolation methods is found in Wise (1998; 2000a,b; 2007). High-resolution DEM measurements such as LiDAR can contain noise not related to how terrain controls the generation or routing of runoff and therefore *non-hydrologic* in character. Nourani and Zanardo (2014) presented a methodology based on a wavelet transformation that filters the noise of high-resolution topography to construct locally smooth topography. After filtering the topography, more realistic saturation zones emerged than when an unfiltered, noisy LiDAR DEM was used to parameterize the slope index of TOPMODEL. Wavelet coefficients were found that effectively discriminated between signal and noise. Application of information content (entropy) was found to be an efficiency metric, from which a critical threshold automatically emerged in the landscape reconstruction. While the technique was applied for TOPMODEL, their methodology could be extended to other complex process models that rely on high-resolution terrain for derivative quantities used in rainfall-runoff modeling.

Other approaches involving the measurement of spatial autocorrelation have been introduced such as the Moran Index. Moran introduced the first measure of spatial autocorrelation in order to study stochastic phenomena, which are distributed in space (Moran 1950). This index is essentially a correlation coefficient evaluated for a group of adjacent or closely spaced data values. The value of Moran's I ranges from +1 meaning a strong positive spatial autocorrelation, to −1 indicating strong negative spatial autocorrelation. A value of 0 indicates a random pattern without spatial autocorrelation. This statistic is designed for application to gridded ordinal, interval or ratio data.

Because spatial data used in distributed hydrologic modeling are generally autocorrelated in space, resampling at different resolutions will affect the mean value of the parameter among other descriptive statistics. Completely random spatial fields, without autocorrelation, can be sampled at almost any resolution without affecting the mean value. Therefore, using the Moran I, information content, or other similar measure aids in understanding the effects of resolution and filtering on slope and other parameters.

4.5 Summary

Changing cell resolution and smoothing of DEMs can have important effects on hydrologic simulation revealing dependence on grid-cell size. We have developed a measure of spatial variability called information content adapted from communication theory. It is easily applied to a map of parameters to gauge the effects of resolution or other filters without distribution fitting. If information content is not lost as we increase cell size, we may be able to use coarser resolution maps, saving computer storage and making simulations more computationally efficient. Application of information content as a measure of spatial variability is demonstrated where derived slope maps become flatter. Selecting a too coarse resolution can be thought of as *cutting* of the hilltops and *filling* of the valleys. Flatter slopes derived from coarse DEMs will produce more delayed and attenuated hydrographs. Care should be taken when using coarse resolution DEMs to recognize that the derived slope maps can have important consequences on hydrologic simulations. Information content may be used to assess the spatial variability of parameters and input maps used to simulate hydrologic processes when scaling up from one cell size to another. This dependence should be recognized as potentially having significant effects on quantities derived for distributed hydrologic modeling, especially derivative parameters such as slope or terrain curvature. When parameter values are dependent on cell size, scaling relationships can be used to estimate more representative parameter values for coarser grid cell resolutions. Using upscaled parameter values can lead to more efficient calibration if the starting parameter values are closer to final values, producing good agreement with observed measurements.

References

Bathurst, J.C., and P.E. O'Connell. 1992. Future of distributed modeling—The System-Hydrologique-Europeen. *Journal of Hydrological Process* 6(3): 265–277.

Beven, K. 1985. Distributed Models. In: *Hydrological Forecasting.* Edited by. Anderson, M.G. and Burt, T.P. 405–435. John Wiley, New York.

Brasington, J., and K. Richards. 1998. Interactions between Model predictions, parameters and DTM scales for TOPMODEL. *Computers and Geosciences* 24(4): 299–314.

Dunne, T., X.C. Zhang, and B.F. Aubry. 1991. Effects of rainfall, vegetation and micro-topography on infiltration and runoff. *Water Resources Research* 27: 2271–2285.

Farajalla, N.S., and B.E. Vieux. 1994. Capturing the essential spatial variability in distributed hydrologic modeling: Hydraulic roughness. *Journal of Hydrological Process* 8(3): 221–236.

Farajalla, N.S., and B.E. Vieux. 1995. Capturing the essential spatial variability in distributed hydrologic modeling: Infiltration parameters. *Journal of Hydrological Process* 9(1): 55–68.

Farmer, J.D., Ott, E. and Yorke, J.A. 1983. The dimension of chaotic attractors. *Physica D: Nonlinear Phenomena*, 7(1–3), 265–278. North-Holland Publishing Co., Amsterdam.

Freeze, R.A., and R.L. Harlan. 1969. Blueprint of a physically-based digitally-simulated hydrologic response model. *Journal of Hydrology* 9: 237–258.

Hengl, T., G.B. Heuvelink, and D.G. Rossiter. 2007. About regression-kriging: from equations to case studies. *Computers & Geosciences* 33(10): 1301–1315.

Jana, R.B., and B.P. Mohanty. 2012. On topographic controls of soil hydraulic parameter scaling at hillslope scales. *Water Resources Research* 48: W02518. doi:10.1029/2011WR011204.

Leavesley, G.H. 1989. Problems of snowmelt runoff modeling for a variety of physiographic climatic conditions. *Hydrological Sciences* 34: 617–634.

Mandelbrot, B.B., 1988, *The Science of Fractal Images.* Edited by. Heinz-Otto, Peitgin and Saupe, Deitmar, 2–21. Springer-Verlag, New York.

Mendoza, P.A., M.P. Clark, M. Barlage, B. Rajagopalan, L. Samaniego, G. Abramowitz, and H. Gupta. 2015. Are we unnecessarily constraining the agility of complex process-based models? *Water Resour. Research* 51: 716–728. doi:10.1002/2014WR015820.

Moran, P.A.P. 1950. Notes on continuous stochastic phenomena. *Biometrika* 37: 17–23.

Niedda, M. 2004. Upscaling hydraulic conductivity by means of entropy of terrain curvature representation. *Water Resources Research* 40: W04206. doi:10.1029/2003WR002721.

Nourani, V., and S. Zanardo. 2014. Wavelet-based regularization of the extracted topographic index from high-resolution topography for hydro-geomorphic applications. *Hydrological Processes* 28: 1345–1357. doi:10.1002/hyp.9665.

Papoulis, A. 1984. *Probability, Random Variables and Stochastic Processes*, 2nd ed, 500–567. New York: McGraw-Hill.

Philip, J.R. 1980. Field heterogeneity. *Water Resources Research* 16(2): 443–448.

Quinn, P., Beven, K., Chevallier, P. and Planchon, O. 1991. The Prediction of hillslope flow paths for distributed hydrological modeling using digital terrain models. In: *Terrain Analysis and Distributed Modeling in Hydrology.* Edited by. Beven, K.J. and Moore, I.D., 63–83. John Wiley and Sons, Chichester, U.K.

Rao, P.S.C., and R.J. Wagenet. 1985. Spatial variability of pesticides in field soils: Methods for data analysis and consequences. *Water Science* 33: 18–24.

Saupe, Deitmar, 1988. *The Science of Fractal Images.* Edited by. Heinz-Otto, Peitgin and Saupe, Deitmar, 82–84. Springer-Verlag, New York.

Seyfried, M.S., and B.P. Wilcox. 1995. Scale and nature of spatial variability: field examples having implications for hydrologic modeling. *Water Resources Research* 31(1): 173–184.

Shannon, C.E. and Weaver, W. 1964. The Mathematical Theory of Communication. The Bell System Technical Journal, University of Illinois Press, Urbana.

Smith, R.E., and R.H.B. Hebbert. 1979. A Monte Carlo analysis of the hydrologic effects of spatial variability of infiltration. *Water Resources Research* 15: 419–429.

Tabler, R.D., Pomeroy, J.W., and Santana, B.W., 1990. Drifting Snow. In: *Cold Regions Hydrology and Hydraulics*. Edited by. Ryan, W.C., Crissman, R.D., Tabler, R.D., Pomeroy, J.W., and Santana, B. W., 95–145.

Tarboton, D.G., Bras, R.L., and Iturbe, I.R., 1991. On the extraction of channel networks from digital elevation data. In: *Terrain Analysis and Distributed Modeling in Hydrology*. Edited by. Beven, K.J. and Moore, I.D., 85–104. John Wiley, New York.

Vieux, B.E. 1993. DEM resampling and smoothing effects on surface runoff modeling. *ASCE, Journal of Computer in Civil Engineering,* Special Issue on Geographic Information Analysis, 7(3):310–338.

Wang, Q., W. Shi, P.M. Atkinson, and Y. Zhao. 2015. Downscaling MODIS images with area-to-point regression kriging. *Remote Sensing of Environment* 166: 191–204.

Wise, S.M. 1998. The effect of GIS interpolation errors on the use of DEMs in geomorphology. In *Landform Monitoring, Modeling and Analysis*, ed. S.N. Lane, K.S. Richards, and J.H. Chandler, 139–164. Chichester: Wiley.

Wise, S.M. 2000a. Assessing the quality for hydrological applications of digital elevation models derived from contours. *Hydrological Processes* 14(11–12): 1909–1929.

Wise, S.M. 2000b. GIS data modelling—lessons from the analysis of DTMs. *International Journal of Geographical Information Science* 14(4): 313–318.

Wise, S.M. 2007. Effect of differing DEM creation methods on the results from a hydrological model. *Computers and Geosciences* 33(10): 1351–1365.

Wise, S. 2011. Cross-validation as a means of investigating DEM interpolation error. *Computers and Geosciences* 37(8): 978–991.

Wise, S. 2012. Information entropy as a measure of DEM quality. *Computers & Geosciences,* Vol. 48, November, pp. 102–110.

Wood, E.F., M. Sivapalan, and K. Beven. 1990. Similarity and scale in catchment storm response. *Review of Geophysics* 28(1): 1–18.

Wood, E.F., 1995. Scaling behaviour of hydrological fluxes and variables: empirical studies using a hydrological model and remote sensing data. *Hydrological Processes* 9(3–4): 331–346.

Chapter 5
Infiltration

Abstract Infiltration rate excess and saturation rate excess runoff are dominant processes in many watersheds. Estimation of infiltration parameters over large areas poses a challenge to distributed watershed modeling. Accuracy may vary depending on how well the soil maps represent the soil and the hydrologic conditions controlling the process across scales from field to watershed. How the soil properties are represented in the distributed model, its configuration as single or multiple layers, can affect the accuracy and reliability of hydrologic prediction. This chapter deals with deriving initial values of infiltration equation parameters from maps of soil properties.

5.1 Introduction

Modeling infiltration in relation to rainfall and antecedent soil moisture takes into account that before ponding of water on the soil surface, the infiltration rate is equivalent to the rainfall rate. After ponding occurs, the infiltration rate is controlled by soil properties, soil depth, degree and depth of saturation, and antecedent soil moisture. Runoff during extreme rainfall exceeding soil infiltration rates produces sufficient runoff to breach a dam (breach is nearly vertical through a clay embankment). As seen in the foreground in Fig. 5.1, sandy sediment, 50 % finer than 0.1 mm, was deposited by runoff and erosion and accumulated within the reservoir of a breached earth embankment in Caddo County Oklahoma. Even prolonged but lower intensity rainfall can produce runoff if there is a limiting depth such as parent rock surfaces or clay accumulation layers impeding downward movement of water in some soils. Once the fillable porosity is saturated above an impeding layer, subsequent rainfall runs off and is known as saturation excess. Soil moisture antecedent to a runoff producing event hastens the initiation of runoff and becomes an important model state demanding continuous simulation to correctly

© Springer Science+Business Media Dordrecht 2016
B.E. Vieux, *Distributed Hydrologic Modeling Using GIS*,
Water Science and Technology Library 74, DOI 10.1007/978-94-024-0930-7_5

Fig. 5.1 A dam failure in sandy soil caused by intense and prolonged precipitation (breach is in background on *right*)

represent a watershed response. Thus, soil properties affecting infiltration rate, the depth of soil above an impeding layer and the evaporation and vegetative transpiration rate all combine to determine how much runoff, if any, will occur for a particular rainfall episode.

Methods for estimating infiltration rates can range from the full partial differential equation, known as Richards' equation, to the more simplified Green and Ampt equation, which is essentially a plug flow of a wetting front without diffusion. The basic idea is to estimate the infiltration rate as controlled by the interplay between rainfall rate and the land surface. This may involve estimating potential infiltration rates or parameters from soil properties or by regionalizing the point estimate obtained by some type of measurement device. In either case, a parameter known at a point must be extrapolated to large areas to be useful in runoff simulations. It would be impractical to take sufficient measurements at each point in the watershed. Thus, we often rely on regionalized variable theory together with some form of soil map to model infiltration at the river basin scale.

There are many issues surrounding the derivation of the governing equation for an unsteady unsaturated flow in a porous media, as first presented by Richards (1931). The originators of the Green and Ampt equation (Green and Ampt 1911) derived a method that predated Richards' equation. Essentially, the Green-Ampt equation ignores diffusion of soil moisture through a range of saturation, considering only an abrupt wetting front (Van den Putte et al. 2013). It is possible to estimate infiltration parameters over large areas using soils maps. The accuracy of such infiltration estimates may depend on how well the soil maps represent the soil and the hydrologic conditions controlling the process. This chapter reviews efforts to estimate infiltration from soil properties. An application follows showing the derivation of Green-Ampt parameters from soil properties. Such infiltration estimates may depend on how well the soil maps represent the soil and the hydrologic conditions controlling the process. This chapter reviews efforts to estimate infiltration from soil properties.

5.2 Approaches to Infiltration Modeling

Accurate infiltration components are essential for physics-based hydrologic modeling. Many current hydrologic models use some form of the Green-Ampt equation to partition rainfall between runoff and infiltration components. As with all models, accurate parameter estimates are required to obtain reliable results. Parameter estimation usually begins with soil properties and is then refined through calibration of the infiltration model, or indirectly, the runoff model to agree with measured data. It is often necessary to generalize from laboratory measurements of infiltration parameters to field conditions. In many field experiments, gaining agreement between model and observations is confounded by factors ranging from formation of macro pores, soil crusting, effects of vegetative cover, or saturation of deeper layers that affect the ability of a simplified model such as Green-Ampt to describe the ensuing complexities.

Modifications that account for factors besides the soil texture in the estimation of infiltration include crusting or gravel and modifications of density due to agricultural practices or other anthropogenic modifications. The interaction between soil crusting and disturbance such as construction or agricultural tillage can explain the large prediction variance (Ahuja 1983). Significant factors affecting hydraulic conductivity and variability in infiltration rates for agricultural soils were found by Risse et al. (1995a, b), namely

1. Soil crusting and tillage
2. Storm event size and intensity
3. Antecedent moisture conditions

The soil properties investigated were based on texture—sand, clay, silt, very fine sand—and on the field capacity, wilting point, organic matter, cation exchange capacity (CEC), and rock fragments. One of the more sensitive parameters in the Green-Ampt equation is saturated hydraulic conductivity. Modifications affecting soil crusting can be made to a baseline value of this parameter, K_b. Separating soils into two classes based on clay content, one predictive relation for greater than 40 % clay had a baseline hydraulic conductivity of

$$K_b = 0.0066 \exp\left(\frac{244}{\% \, \text{clay}}\right) \qquad (5.1)$$

For soils with less than 40 % clay, the relationship for K_b incorporates two parameters for sand content and replaces percent clay with CEC as

$$K_b = -0.265 + 0.0086 \, \% \, \text{sand}^{1.8} + 11.46 \, \text{CEC}^{-0.75}, \qquad (5.2)$$

where CEC has units of meq/100 g. Overall, for the soils evaluated a good agreement of K_b was found with a coefficient of determination of $R^2 = 0.78$. From Eq. (5.2), it is apparent that the sandy soils have higher baseline conductivity as would be expected. However, the baseline conductivity must be modified to take into account other factors controlling infiltration such as the soil-crust interface (Risse et al. 1995a, b).

Corradini et al. (1994, 1997) extended an earlier model developed by Smith et al. (1993) in order to describe infiltration using multiple layers experiencing infiltration during complex rainfall. This model addressed a range of complex cases characterized where a series of infiltration and moisture redistribution cycles exist. The basic model considered the problem of point infiltration during a storm consisting in two parts separated by a rainfall hiatus, with surface saturation and runoff occurring in each part. The model employed a two-part profile for simulating the actual soil profile. When the surface flux was not at capacity, it used a slightly modified version of the Parlange et al. (1985) model for description of surface water content and redistribution. Criteria for the development of compound profiles and for their reduction to single profiles were also incorporated. The extended model was tested by comparison with numerical solutions of Richards' equation, carried out for a variety of experiments upon two contrasting soils. Venkata Reddy et al. (2007) demonstrated successful model prediction using an adaptation of the Green and Ampt equation to account for time to ponding, called Green-Ampt-Mein-Larson (GAML). Reasonable success was gained from the GAML approach through calibration. Chahinian et al. (2005) evaluated a model simulation of Hortonian overland flow at the field or small-plot scale using the Philip (1957), Morel-Seytoux, Horton, and SCS infiltration routines. The first three infiltration models gave similar and adequate results when initial conditions were calculated from measured surface water content. However, the SCS infiltration model performed poorly considering its ability to simulate the influence of time-variable rainfall with a hiatus. As with many models, the main issue is parameterization and *equifinality* of parameters that may be difficult to discern with often limited observations (Beven and Binley 1992).

Infiltration properties have been shown to vary both spatially and temporally because of soil property variation, cultural influences of man, activities of flora and fauna and the interaction of the soil surface with kinetic energy of rainfall.

Variability in a physical property can often result from an investigation at a scale inappropriate for the problem at hand. Sisson and Wierenga (1981) showed that the variability of the infiltration properties tended to decrease with increased sampling area and volume. Therefore, scaling up to watershed areas will likely require some adjustment of the infiltration parameters.

Govindaraju et al. (2012) found that vertical and horizontal heterogeneity affects short- and long-term cumulative infiltration differently. Surface horizontal heterogeneity is found to control field-scale infiltration at small times, whereas local vertical nonuniformity exerts a strong control at long times. The vertical non-uniformity manifests itself by ponding even where rainfall rates do not exceed surface infiltration rates, i.e., an impeding layer effectively causes water to pond at the surface. Both horizontal and vertical nonuniformities affect the scaling of hydrologic processes from point measurements to plots, fields, and watersheds. Corradini et al. (2011) tested a conceptual model for estimating field-scale infiltration using a two-layered soil model. The upper layer is considered to have a saturated hydraulic conductivity much greater than the lower impeding layer. The upper layer is considered to be a horizontal random field while the lower conductivity in the subsoil is assumed to be constant. The multi-layer infiltration equations showed promise for modeling both infiltration rates and cumulative infiltration at the field scale. Govindaraju et al. (2001) also evaluated the influence of spatial correlations of infiltration parameters on areal averaging at the field scale. Their simulations found that field-scale infiltration could be effectively represented by a scaling relationship based on the correlation length of saturated hydraulic conductivity.

It is well recognized that spatial heterogeneities exist and can affect the resulting infiltration rates and volumes that depend on the scale of measurement (Williams and Bonell 1988). Spatially variable infiltration parameters consist in both deterministic and stochastic components and should be explicitly recognized. Deterministic variability is defined by soil types of known or estimated properties. Stochastic variability recognizes the variance of properties within the soil mapping unit due to unknown variability in soil properties.

Hydrological data are neither purely deterministic nor purely stochastic in nature but are a combination of both. Gupta et al. (1994) analyzed spatially variable infiltration parameters by decomposing them into deterministic and stochastic components. Hydraulic conductivity exhibits the maximum degree of variation when compared to the sorptivity and infiltration rate of the soil described by the Philip equation. A model involving deterministic and stochastically dependent and independent subcomponents can thus describe the spatial patterns of each infiltration parameter. The deterministic component of each model can be expressed by a two-harmonic Fourier Series with the dependent stochastic subcomponent described by a second-order autoregression model and a residual sub-component.

The standard assumption implicit in the Green-Ampt formulation is that the moisture front infiltrating into a semi-infinite, homogeneous soil at uniform initial volumetric water content occurs without diffusion. Salvucci and Entekhabi (1994) derived explicit expressions for the Green-Ampt (delta diffusivity) infiltration rate and cumulative storage. Through simple integration, the wetting front model of infiltration yields an exact solution relating the infiltration rate, cumulative infiltration and time. The proposed expression avoids the nonlinearity of the Green-Ampt equation, thus reducing computational time.

Capturing the spatial variation in soil hydrological behavior relevant to watershed scales is an ongoing topic in hydrology. From field sampling, the majority of the soil properties have short length scales, from 100 to 200 m as described by Jetten et al. (1993). Watershed modeling does not usually have the benefit of such detailed sampling, yet estimates of infiltration parameters are needed at larger scales than are envisioned within field sampling studies. Besides soil properties, vegetation that is part of the soil-plant-atmosphere system affects infiltration and resulting soil moisture plays an important role in determining the watershed response.

Representing the infiltration process over a river basin can be approached directly by infiltration measurements or indirectly by using a model that relies on parameter estimates based on soil properties. Both approaches contain errors caused by spatial variation of soil properties. Infiltration models show a variance that also originates from spatial variance of the soil properties. However, in a model approach, this original spatial variance is altered through the combination and processing of the parameters in the algorithms used in the model. The spatial variation not captured in a soil map or detected through measurement is one of the major sources of variance encountered in infiltration simulation at the watershed scale.

As discussed in Chap. 3, kriging has been applied to estimating soil properties across some spatial extent. The regionalized variable theory (Burrough 1986; Webster and Oliver 1990; Davis 1973) assumes that a spatial variation of a variable $z(x)$ can be expressed as the sum of three components

$$z(x) = m(x) + \varepsilon'(x) + \varepsilon'', \tag{5.3}$$

where x is the position given in two or three coordinates; $m(x)$ = structural component, i.e., a trend or a mean of an area; $\varepsilon'(x)$ = stochastic component, i.e., spatially correlated random variation; ε'' = residual error component (noise), i.e., spatially uncorrelated random variation. Spatial analysis of a variable involves the detection and subsequent removal of trends $m(x)$ in the data set, after which the spatially correlated random variation can be described. This is a departure from the approach of using a mapping unit and associated soil properties to estimate Green-Ampt infiltration. To apply the regionalized variable theory, we must have point estimates of the variate at sufficient density and spatial extent. This method suffers from the same obstacle as infiltration measurement; sufficient measurements are difficult to obtain over large regions. This makes direct measurement and regionalized variable approaches difficult to apply from lack of data over large areas such as river basins.

5.3 Green-Ampt Infiltration

This section describes the Green-Ampt infiltration equation and the parameters that control the infiltration rate excess runoff process. Once the variables are defined, then regression equations can be formulated to predict values based on soil properties mapped over the river basin. The rate form of the Green-Ampt equation for the one-stage case of initially ponded conditions and assuming a shallow ponded water depth is

$$f(t) = K_e \left[1 + \frac{\Psi_f \Delta \theta}{F(t)} \right], \tag{5.4}$$

where $f(t)$ is the instantaneous infiltration rate with dimension (L/T); K_e is effective saturated conductivity (with dimension, L/T), usually taken as 50 % of predicted soil properties for reduction due to entrapped air; Ψ_f is average capillary potential or wetting front suction (with dimension, L); $\Delta \theta$ is the moisture deficit (with dimension, L/L); and $F(t)$ cumulative infiltration depth (with dimension, L). The soil moisture deficit is computed as

$$\Delta \theta = \phi_{total} - \theta_i, \tag{5.5}$$

where ϕ_{total} is the total porosity (with dimension, L/L); and θ_i is initial volumetric water content (with dimension, L/L). Equation (5.4) may be reformulated for ease of solution as

$$K_e t = F(t) - \Psi_f \Delta \theta \ln \left[1 + \frac{F(t)}{\Psi_f \Delta \theta} \right], \tag{5.6}$$

where t is time, with the other terms are as defined above. Equation (5.6) can be solved for cumulative infiltration depth, F for successive increments of time using a Newton–Raphson iteration together with Eq. (5.4) to obtain the instantaneous infiltration rate.

For the case of constant rainfall, infiltration is a two-stage process. The first stage occurs when the rainfall is less than the potential infiltration rate and the second when the rainfall rate is greater than the potential infiltration rate. Mein and Larson (1973) modified the Green-Ampt equation for this two-stage case by computing a time to ponding, t_p, as

$$t_p = \frac{K_e \Psi_f \Delta \theta}{i(i - K_e)}, \tag{5.7}$$

where i = rainfall rate. The actual infiltration rate after ponding is obtained by constructing a curve of potential infiltration beginning at time t_0 such that the cumulative infiltration and the infiltration rate at t_p are equal to those observed under rainfall beginning at time 0. The time t is computed as

$$t = t_p - t_0 \tag{5.8}$$

The time to ponding, t_p depends on soil properties and initial degree of saturation expressed as a soil moisture deficit, $\Delta\theta$ and is computed as

$$t_p = \frac{F - \Psi_f \Delta\theta \ \ln(1 + \frac{F}{\Psi_f \Delta\theta})}{K_e}, \tag{5.9}$$

where F = cumulative infiltration depth (L) at the time to ponding, which, for the case of constant rainfall, is equal to the cumulative rainfall depth at the time to ponding. Before ponding, all rainfall is considered to be equal to the infiltrate. The corrected time as computed by Eq. (5.9) is used in Eq. (5.6), which is solved as with the one-stage case. Computing cumulative and instantaneous infiltration requires soil parameters, which in turn must be estimated from mapped soil properties.

5.3.1 Parameter Estimation

Infiltration parameter estimation consists in: (1) the Green-Ampt equation and the parameters used to estimate infiltration; (2) estimation of the Green-Ampt parameters from soil properties; and (3) application of soil maps to derive distributed infiltration parameters. A useful description of soil properties is contained in the digital soils database called SSURGO, compiled by the USDA-Natural Resources Conservation Service (NRCS) from county-level soil surveys. STATSGO is a generalized soils data base and has soil maps in GIS format compiled for each state in the US. It is generalized in the sense that the STATSGO map units are aggregated from soils of similar origin and characteristics. SSURGO maps were compiled at a scale of 1:20,000, while STATSGO soil map units were compiled at a smaller scale of 1:250,000. Thus, much of the detail present in county-level SSURGO soil maps may be lost at the 1:250,000 STATSGO scale and generalization. These soil maps together with a data base of soil properties comprise the most detailed soils information available over any large spatial extent in the US.

Each mapping unit in the USDA soil data base contains estimated and measured data on the physical and chemical soil properties and soil interpretations for engineering, water management, recreation, agronomic, woodland, range, and wildlife uses of the soil. A method for estimating Green-Ampt parameters from soil texture was reported by Rawls et al. (1983a, b) and updated by Saxton and Willey (2005) and Saxton and Rawls (2006). In estimating the soil model parameters from soil properties, the infiltration parameters are effective porosity, θ_e; wetting front suction head, ψ_f; and saturated hydraulic conductivity, K_{sat}. These parameters are based on the following equations. The effective porosity, θ_e is

$$\theta_e = \phi_{\text{total}} - \theta_r, \tag{5.10}$$

where ϕ_{total} is total porosity, or 1—(bulk density)/2.65; and θ_r is residual soil moisture content (cm³/cm³). The effective porosity, θ_e is considered a soil property and is independent of soil moisture at any particular time, as is the wetting front suction head. Using the Brooks and Corey (1964) soil water retention parameters to simulate the soil water content as a function of suction head, the wetting front suction head, Ψ_f (cm), is given as

$$\Psi_f = \frac{2 + 3\lambda}{1 + 3\lambda} \frac{\Psi_b}{2}, \tag{5.11}$$

where ψ_b is bubbling pressure (cm) and λ is the pore size distribution index. Saturated hydraulic conductivity, K_s in cm/h, is estimated from the effective porosity, bubbling pressure, and pore size distribution index as

$$K_s = 21.0 * \frac{(\theta_e/\Psi_b)^2(\lambda)^2}{(\lambda+1)(\lambda+2)}, \tag{5.12}$$

Equations (5.10–5.12) allow the expression of the Green-Ampt parameters, wetting front suction and hydraulic conductivity in terms of the soil water retention parameters, pore size index, bubbling pressure and effective porosity (Brooks and Corey 1964). Soil water retention parameters were estimated from soil properties using regression techniques by Rawls et al. (1983a, b) as follows:

Bubbling Pressure

$$\begin{aligned}
\psi_b = \exp[&5.3396738 + 0.1845038(C) - 2.48394546\,(\phi_{\text{total}}) \\
&- 0.00213853(C)^2 - 0.04356349(S)(\phi_{\text{total}}) \\
&- 0.61745089(C)(\phi_{\text{total}}) + 0.00143598(S)^2(\phi_{\text{total}})^2 \\
&- 0.00855375(C)^2(\phi_{\text{total}})^2 - 0.00001282(S)^2(C) \\
&+ 0.00895359(C)^2(\phi_{\text{total}}) - 0.00072472(S)^2(\phi_{\text{total}}) \\
&+ 0.0000054(C)^2(S) + 0.50028060(\phi_{\text{total}})^2(C)]
\end{aligned} \tag{5.13}$$

Pore Size Distribution Index

$$\begin{aligned}
\lambda = \exp[&-0.7842831 + 0.0177544(S) - 1.062498(\phi_{\text{total}}) \\
&- 0.00005304(S)^2 - 0.00273493(C)^2 + 1.11134946(\phi_{\text{total}})^2 \\
&- 0.03088295(S)(\phi_{\text{total}}) + 0.00026587(S)^2(\phi_{\text{total}})^2 \\
&- 0.00610522(C)^2(\phi_{\text{total}})^2 - 0.00000235(S)^2(C) \\
&+ 0.00798746(C)^2(\phi_{\text{total}}) - 0.00674491(\phi_{\text{total}})^2(C)]
\end{aligned} \tag{5.14}$$

Residual Soil Moisture Content

$$\theta_r = -0.0182482 + 0.00087269(S) + 0.00513488(C)$$
$$+ 0.02939286(\phi_{total}) - 0.00015395(C)^2 - 0.0010827(S)(\phi_{total})$$
$$- 0.00018233(C)^2(\phi_{total})^2 + 0.00030703(C)^2(\phi_{total}) - 0.0023584(\phi_{total})^2(C)$$

$$(5.15)$$

where C is percent of clay in the range 5 % < C < 60 %; S is percent of sand in the range 5 % < S < 70 %; and ϕ_{total}. Given a map of soil properties, clay, sand and total porosity, we can thus estimate the Green-Ampt parameters for the mapped soil type.

The relationships expressed by Eqs. (5.13–5.15) may not always produce reasonable values of Green-Ampt parameters for some soils, especially if the soil properties fall outside of the range used to derive Eqs. (5.13–5.15) and when clay content is above 50 %. It is useful to compare the predicted values with those published by Rawls et al. (1983a, b) for the range of soil classes. Table 5.1 presents the mean parameter values expected for USDA soil textural classifications. The range is shown for wetting front suction (3 col.), which exhibits considerable variability. Reclassification of soil mapping units into corresponding parameters can yield results equivalent to the regression equations described above. When the percent of clay of a particular soil exceeds the limits used to derive the regression equation in the first place, unexpected infiltration rates or parameter values can result (personal correspondence with W.J. Rawls).

For hydrologic purposes, the rooting zone depth may be a better descriptor of saturation excess. In mountainous terrain, the soil depth can be controlled by

Table 5.1 Green and Ampt infiltration parameter Green and Ampt infiltration parameters estimates based on soil texture

Texture	θ_ε	ψ (cm)	K_s (cm/h)
Sand	0.437	4.95 (0.97–25.36)	11.78
Loamy sand	0.437	6.13 (1.35–27.94)	2.99
Sandy loam	0.453	11.01 (2.67–45.47)	1.09
Loam	0.463	8.89 (1.33–59.38)	0.34
Silt loam	0.501	16.68 (2.92–95.39)	0.65
Sandy clay loam	0.398	21.85 (4.42–108.0)	0.15
Clay loam	0.464	20.88 (4.79–91.10)	0.1
Silty clay loam	0.471	27.3 (5.67–131.50)	0.1
Sandy clay	0.43	23.9 (4.08–140.2)	0.06
Silty clay	0.479	29.22 (6.13–139.4)	0.05
Clay	0.475	31.63 (6.39–156.5)	0.03

Adapted from Rawls et al. (1983a)

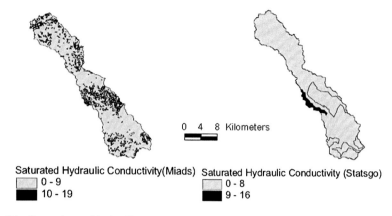

Saturated Hydraulic Conductivity(Miads)   Saturated Hydraulic Conductivity (Statsgo)
░ 0 - 9   ░ 0 - 8
■ 10 - 19   ■ 9 - 16

Fig. 5.2 Comparison of hydraulic conductivity derived from SSURGO and STATSGO

topographical sequence, i.e., depth increases from upslope to downslope due to deposition. To obtain an improved map of soil depth in an arid mountainous watershed, Tefsa et al. (2011) used the DEM and position within the topographic sequence (upslope to downslope) to help refine soil depth estimated from STATSGO for purposes of infiltration modeling.

The scale at which a soils map is compiled affects the variability of the resulting parameter map. Figure 5.2 is a comparison of saturated hydraulic conductivity from SSURGO (left) and STATSGO (right) soil data for the 1,200 km^2 Blue River Basin, located in south central Oklahoma. The parameter histogram in Fig. 5.3 indicates loss of several categories as evidenced by almost no spatial variability in the STATSGO map compared with SSURGO. The mean value for saturated hydraulic conductivity was calculated to be 0.23 cm/h for SSURGO and 1.5 cm/h for STATSGO data. Thus, derivation of Green and Ampt parameters from a

Fig. 5.3 Histogram of hydraulic conductivity derived from detailed and generalized soils maps

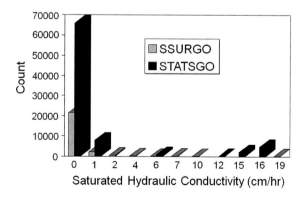

detailed map can differ significantly when using the detailed soils maps (SSURGO) compared with a generalized soil map (STATSGO) due to aggregation of soil mapping units.

5.3.2 Parameter Uncertainty

Considering the uncertainty introduced by the estimated range of soil properties, estimated values will likely require adjustment. Because there is a range of soil properties within each soil mapping unit, there are also uncertainties in the Green-Ampt parameters estimated for the soil mapping unit from these properties using Eqs. (5.13–5.15). Few distributed environmental models account for the quality and spatial nature of input data (Aspinall and Pearson 1993). Studies by Buttenfield (1993), Fisher (1993), and Lunetta et al. (1991) addressed the issue of quantifying data errors (attribute, positional, or temporal) but not the effects of these errors on simulation results. Recently, some efforts have been directed to modeling error propagation in GIS-integrated models. These include the use of statistical theory in error analysis by Burrough (1986); standard error or Taylor series expansion to determine error propagation in GIS-integrated environmental modeling described by Heuvelink et al. (1990) and Wesseling and Heuvelink (1993); and error propagation due to land use misclassification (errors of omission and commission) described by Veregin (1994). A Bayesian probabilistic framework applied to infiltration modeling by Yen et al. 2014 and Ajami et al. (2007) demonstrated that not all model uncertainty is due to parameters but also model structure.

A thematic map consists in spatial features (polygons, lines, or points) and attributes (e.g., soil type, land use, population) associated with each spatial feature. Soil surveys usually provide soil properties as a range for a particular mapping unit. Larger mapping units are more likely to have a greater range of values due to the inherent variability of soil. Infiltration parameters derived from soil texture are uncertain due to the ranges of soil properties in the soil attribute data base for a given mapping unit. Soil mapping units are mapped and assembled in digital format in the USDA soil data base (Web Soil Survey 2016). For each mapping unit, soil properties are tabulated including texture as percent of sand, silt and clay; physical properties of bulk density and saturated hydraulic conductivity (KSAT). A subset of physical properties is presented in Table 5.2 for three of the more significant mapping units, MUSYM 5, 35, and 123, comprising 2, 10, and 11 % of the watershed area, respectively. The Ratake-Cathedral Rock outcrop (MUSYM 123) is representative of the upland area in the foothills, which initially is estimated at 8.38 cm/h, an order of magnitude higher than the Denver-Urban land complex (MUSYM 35) at 0.84 cm/h. A limiting depth to water table, restrictive layer, or any other reason, can be useful for assigning soil depth for modeling saturation

Table 5.2 Summary of physical properties for selected map units (Web Soil Survey 2016)

Map unit symbol	Map unit name	KSAT (cm/h)	Limit depth (cm)
5	Argiustolls-Rock outcrop complex—stony sandy loam	8.38	89
35	Denver-Urban land complex—clay loam	0.84	>200
123	Ratake-Cathedral-Rock outcrop complex stony sandy loam	8.38	30

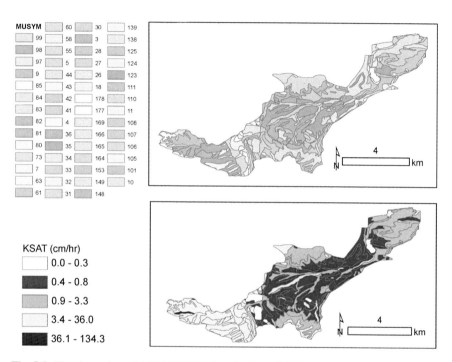

Fig. 5.4 Mapping units and initial KSAT values for Lena Gulch

excess, seen in the last col. of Table 5.2, *Limit depth*. Mapping units within Lena Gulch are seen in Fig. 5.4, with initial estimates of saturated hydraulic conductivity. Calibration results for soil depth (limiting) is shown in Fig. 5.5. A marked decrease is evident in the upper (lower left) portion of the basin where the range in KSAT initially ranged from 36.1 to 134.3 cm/h but through calibration with an observed streamflow, was reduced to 7.15–10.16 cm/h, effectively a 61 % reduction on average over the basin. A similar reduction was witnessed for soil depth, with a 56 % reduction (not shown) from 134 to 62 cm averaged over the basin.

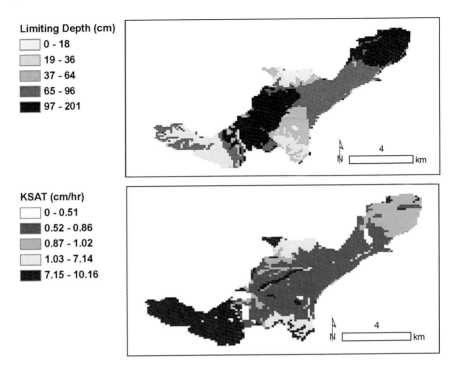

Fig. 5.5 Calibrated results for soil depth and KSAT

5.4 Summary

Infiltration is a major hydrologic process controlling the amount of runoff at scales ranging from hillslopes to river basins. Measurements of infiltration and soil characteristics are usually done at point locations. Estimating infiltration from soil maps may be the only feasible alternative to making extensive measurements. In order to gain some idea of the spatial distribution of soil characteristics and infiltration over large areas, we must resort to remote sensing, geologic maps, or soil maps to estimate infiltration. The amount of spatial detail in a soil map within a watershed has important consequences on how well the simulated hydrologic response agrees with observed flow. Application of the Green-Ampt equations in watershed modeling requires spatially variable parameters that are representative of soil properties controlling the infiltration process. Because the mapped soil properties are estimated for each soil mapping unit, variation is expected leading to uncertainty in the parameter values. However, the relative spatial organization of high/low infiltration rates is an important starting point for distributed model calibration as discussed later in Chap. 10.

References

Ahuja, L.R. 1983. Modeling infiltration into crusted soils by the Green-Ampt approach. *Soil Science Society of America Journal* 47(3): 412–418.

Ajami, N.K., Q. Duan, S. Sorooshian. 2007. An integrated hydrologic Bayesian multimodel combination framework: confronting input, parameter, and model structural uncertainty in hydrologic prediction. Water Resource Research 43(1). http://dx.doi.org/10.1029/ 2005WR004745.

Aspinall, R.J. and D.M. Pearson. 1993. Data quality and spatial analysis: analytical use of GIS for ecological modeling. *Proceedings of the Second International Conference on Integrated Geographic Information Systems and Environmental Modeling,* Sept 26–30, 1993, Breckenridge, Colorado.

Beven, K., and A. Binley. 1992. The future of distributed models: Model calibration and uncertainty prediction. *Hydrological Processes* 6(3): 279–298.

Brooks, R.H. and A.T. Corey. 1964. Hydraulic properties of porous media. *Hydrology Paper No. 3.* Fort Collins, Colorado: Colorado State University.

Burrough, P.A. 1986. Principles of geographic information systems for land resources assessment. *Monographs on soil and resources survey,* No. 12, 103–135. Oxford Science Publications.

Buttenfield, B. 1993. Representing data quality. *Cartographica* 30(2–3): 1–7.

Chow, V.T., D.R. Maidment, and L.W. Mays. 1988. *Applied hydrology.* New York: McGraw-Hill.

Chahinian, N., R. Moussa, P. Andrieux, and M. Voltz. 2005. Comparison of infiltration models to simulate flood events at the field scale. *Journal of Hydrology* 306(1): 191–214.

Corradini, C., F. Melone and R.E. Smith. 1994. Modeling infiltration during complex rainfall sequences. *Water Resources Research* 30(10): 2777–2784.

Corradini, C., F. Melone, and R.E. Smith. 1997. A unified model for infiltration and redistribution during complex rainfall patterns. *Journal of Hydrology* 192(1–4): 104–124.

Corradini, C., A. Flammini, R. Morbidelli, and R.S. Govindaraju. 2011. A conceptual model for infiltration in two-layered soils with a more permeable upper layer: From local to field scale. *Journal of Hydrology* 410(1–2): 62–72.

Davis, J.C. 1973. *Statistics and Data analysis in geology.* New York: John Wiley and Sons. 71p.

Fisher, P. 1993. Visualizing uncertainty in soil maps by animation. *Cartographica* 30(2–3): 20–27.

Govindaraju, R.S., R. Morbidelli, and C. Corradini. 2001. Areal infiltration modeling over soils with spatially correlated hydraulic conductivities. *Journal of Hydrology Engineering* 6(2): 150–158.

Govindaraju, R.S., C. Corradini, and R. Morbidelli. 2012. Local- and field-scale infiltration into vertically non-uniform soils with spatially-variable surface hydraulic conductivities. *Hydrological Processes* 26(21): 3293–3301.

Green, W.H., and G.A. Ampt. 1911. Studies in soil physics I: The flow of air and water through soils. *Journal of Agric Science* 4: 1–24.

Gupta, R.K., R.P. Rudra, W.T. Dickinson, and D.E. Elrick. 1994. Modeling spatial patterns of three infiltration parameters. *Canadian Agricultural Engineering* 36(1): 9–13.

Heuvelink, G.B.M, Burrough, P.A., Leenaers. 1990. *Error propagation in spatial modeling with GIS* Ed. Harts et al. 453–462.

Jetten, V.G., H.T.H. Riezebos, F. Hoefsloot, and J. Van Rossum. 1993. Spatial variability of infiltration and related properties of tropical soils. *Earth Surface Processes and Landforms* 18: 477–488.

Koren, V.I., M. Smith, D. Wang, Z. Zhang, 2000. Use of soil property data in the derivation of conceptual rainfall-runoff model parameters, 15th Conference on Hydrology, Long Beach, California, pp. 103–106.

Koren, V.I., Smith, M., Q. Duan. 2003. Use of a priori parameter estimates in the derivation of spatially consistent parameter sets of rainfall-runoff models. In: *Calibration of Watershed*

Models, Water Science and Application. Ed. Duan, Q., Gupta, H.V., Sorooshian, S., Rousseau, A.N., Turcotte, R. Vol. 6, 239–254. American Geophysical Union.

Lunetta, R.S., R.G. Congalton, J.R. Jensen, K.C. McGwire, and L.R. Tinney. 1991. Remote sensing and geographic information system data integration: Error sources and research issues. *Photogrammetric Engineering and Remote Sensing* 57(6): 677–687.

Mein, R.G., and C.L. Larson. 1973. Modeling infiltration during steady rains. *Water Resources Research* 9(2): 384–394.

Parlange, J.Y., W.L. Hogarth, J.F. Boulier, J. Touma, R. Haverkamp, and G. Vauchad. 1985. Flux and water content relation at the soil surface. *Soil Science Society of America Journal* 49(2): 285–288.

Philip, J.R. 1957. The theory of infiltration 4. Sorptivity and algebraic infiltration equations. *Soil Science* 84: 257–264.

Rawls, W.J., D.L. Brakensiek, and N. Miller. 1983a. Green and Ampt infiltration parameters from soils data. *ASCE, Journal of Hydraulic Engineering* 109(1): 62–70.

Rawls, W.J., D.L. Brakensiek, and B. Soni. 1983b. Agricultural management effects on soil water processes, Part I: Soil water retention and Green and Ampt infiltration parameters. *Transaction of the American Society of Agricultural Engineers* 26(6): 1747–1752.

Richards, L.A. 1931. Capillary conduction of liquids through porous mediums. *Physics* I: 318–333.

Risse, L.M., B.Y. Liu, and M.A. Nearing. 1995a. Using curve numbers to determine baseline values of Green-Ampt effective hydraulic conductivities. *Water Resources Bulletin* 31(1): 147–158.

Risse, L.M., M.A. Nearing, and X.C. Zhang. 1995b. Variability in Green-Ampt effective hydraulic conductivity under fallow conditions. *Journal of Hydrology* 169: 1–24.

Salvucci, G.D., and D. Entekhabi. 1994. Explicit expressions for Green-Ampt (delta function diffusivity) infiltration rate and cumulative storage. *Water Resources Research* 30(9): 2661–2663.

Saxton, K.E., and W.J. Rawls. 2006. Soil water characteristic estimates by texture and organic matter for hydrologic solutions. *Soil Science Society of America Journal* 70(5): 1569–1578.

Saxton, K.E. and P.H. Willey. 2005. Soil water characteristics hydraulic properties calculator. *Agricultural Research Service*.

Sisson, J.B., and P.J. Wieranga. 1981. Spatial variability in Green-Ampt effective hydraulic conductivity under fallow conditions. *Soil Science Society of America Journal* 46: 20–26.

Smith, R.E., C. Corradini, and F. Melone. 1993. Modeling infiltration for multistorm runoff events. *Water Resources Research* 29(1): 133–143.

Tefsa, T.K., D.G. Tarboton, D.W. Watson, K.A. Schreuders, M.E. Baker, and R.M. Wallace. 2011. Extraction of hydrological proximity measures from DEMs using parallel processing. *Environmental Modelling & Software*, 26(12): 1696–1709.

Van den Putte, A., G. Govers, A. Leys, C. Langhans, W. Clymans, and J. Diels. 2013. Estimating the parameters of the Green-Ampt infiltration equation from rainfall simulation data: Why simpler is better. *Journal of Hydrology* 476: 332–344.

Venkata Reddy, K., T.I. Eldho, E.P. Rao, and N. Hengade. 2007. A kinematic-wave-based distributed watershed model using FEM, GIS and remotely sensed data. *Hydrological Processes* 21(20): 2765–2777.

Veregin, H. 1994. Integration of simulation modeling and error propagation for the buffer operation in GIS. *Photogrammetric Engineering and Remote Sensing* 60(4): 427–835.

Webster, R. and MA. Oliver, 1990. Statistical methods in soil and land resource survey. In: *Spatial Information systems*, 316 pp. Oxford: Oxford University Press.

Web Soil Survey. 2016. United States Department of Agriculture, Natural Resources Conservation Service. Accessed, March 3016. http://websoilsurvey.sc.egov.usda.gov/App/WebSoilSurvey.aspx.

Wesseling, C.G. and G.B.M. Heuvelink. 1993. *ADAM User's manual.* Department of Physical Geography, University of Utrecht.

Williams, J., and M. Bonell. 1988. The influence of scale of measurement of the spatial and temporal variability of the philip infiltration parameters—An experimental study in an australian savannah woodland. *Journal of Hydrology* 104: 33–51.

Yen, H., Xiuying Wang, Darrell G. Fontane, R. Daren Harmel, and Mazdak Arabi. 2014. A framework for propagation of uncertainty contributed by parameterization, input data, model structure, and calibration/validation data in watershed modeling. *Environmental Modelling and Software* 54: 211–221.

Chapter 6
Hydraulic Roughness

Land Use and Cover

Abstract A map of land use/cover (LULC) can be the starting point for producing a parameter map of overland hydraulic roughness. However, interpretation is important since land use/cover categories must be interpreted to yield a hydraulic parameter that governs the relationship between runoff depth and velocity. Despite what the map classification indicates, a more important aspect is the hydraulic roughness that runoff experiences as it flows over the land surface typical of a particular land use/cover classification. Modeling surface runoff assumes that the derived hydraulic parameters are representative of the land use/cover, at least as a starting point for calibration. Channel roughness will be seen to differ markedly from overland flow roughness coefficients and should be assigned based on channel characteristics rather than LULC. Deriving hydraulic roughness coefficients from land use classification schemes is the theme of this chapter.

6.1 Introduction

Hydraulic roughness coefficients are used to predict surface runoff from channel and overland flow areas in watersheds. Calculating the time of concentration, determining flow velocity, and simulating runoff hydrographs require the use of hydraulic roughness coefficients (Gilley and Finkner 1991). The velocity of surface runoff is controlled by the hydraulic roughness. We can distinguish two types of runoff: overland flow and channel. Of course, this distinction is somewhat artificial since the scale of the basin and runoff processes often dictate whether we represent small ravines or drainage swales explicitly as channels. From a modeling perspective, the difficulty in assigning channel characteristics from the smallest ravine to the largest river channel within the basin of interest can be daunting. Model abstraction leads to assumptions that overland flow is present even where stream channels may exist. Such abstraction in model representation of reality undoubtedly introduces variance and uncertainty in the modeled response. Figure 6.1 shows an example where grass understory in an orchard is more important for modeling the runoff from this location than the land use classification of 'Orchard.'

© Springer Science+Business Media Dordrecht 2016
B.E. Vieux, *Distributed Hydrologic Modeling Using GIS*,
Water Science and Technology Library 74, DOI 10.1007/978-94-024-0930-7_6

Fig. 6.1 A land use/cover map categorizes this location as an 'Orchard', however, the hydraulic roughness of overland flow will be influenced by the grass understory, more so than by the trees

  To understand and properly account for overland and channel flow hydraulics, we must understand the hydraulics governing flow over natural surfaces. Frictional drag over the soil surface, standing vegetative material, crop residue and rocks lying on the surface, raindrop impact, and other factors may influence resistance to flow on upland areas. Hydraulic roughness coefficients caused by each of these factors contribute to total hydraulic resistance. We may consider the hydraulic roughness to be a property assigned to a land use/cover classification. We can derive hydraulic roughness maps from a variety of sources, including aerial photography, generally available land use/cover maps and remote sensing of vegetative cover. Each of these sources lets us establish hydraulic roughness over broad areas such as river basins. Detailed measurement of hydraulic roughness over any large spatial extent is impractical. Thus, reclassifying a GIS map of land use/cover into a map of hydraulic roughness parameters is attractive in spite of the errors present in such an operation. The goal is to represent spatially the location of hydraulically rough versus smooth land use types. To this end, we first consider the hydraulics of overland and channel flow. We then present examples of reclassifying generally available land use/cover maps into hydraulic roughness. It should be noted at the outset that the values of hydraulic roughness assigned to each land use/cover classification are somewhat speculative. Here is where the hydrologist must exercise some judgment as to how runoff responds to the roughness of the natural surface as classified in the map.

6.2 Hydraulics of Surface Runoff

Hydraulics of flow over natural surfaces is based on traditional fluid mechanics and has been well known for some time as discussed in Chow (1988). The equations that relate flow depth and velocity are quite simple in form. The difficulty lies in establishing appropriate values of the roughness parameter for a particular land use or surface condition. The Darcy-Weisbach and Manning equations have been widely used to describe flow characteristics. Both relations contain a hydraulic roughness coefficient. Under uniform flow conditions, the Darcy-Weisbach hydraulic roughness coefficient, f, is given as:

$$f = \frac{8gRS}{V^2} \tag{6.1}$$

where g is acceleration due to gravity; S is average slope; V is the flow velocity averaged with depth, and R is the Reynolds number. In metric units, the Manning hydraulic roughness coefficient, n, is given as:

$$n = \frac{R^{2/3}S^{1/2}}{V} \tag{6.2}$$

Manning, n and Darcy-Weisbach, f, hydraulic roughness coefficients are related by the following equation:

$$n = \left[\frac{fR^{1/3}}{8g}\right]^{1/2} \tag{6.3}$$

Chézy attempted to relate the flow depth in a canal to its velocity. Since then other relationships have been defined that rely on defining hydraulic roughness coefficients that are characteristic of the surface over which water flows. Engelund and Hansen (1967) gave Chézy roughness coefficient expressions as follows:

$$C = 18 \log\left(\frac{12R}{K_s}\right) \tag{6.4}$$

$$C = KR^{1/6} \tag{6.5}$$

$$C = C_{90}\left(\frac{\theta'}{\theta}\right) \tag{6.6}$$

where C is the Chézy roughness coefficient; R is the Reynolds Number; K_{is} the effective roughness; 2 and 2τ are the shield parameters; and C_{90} is the roughness according to the grain size composing the surface. The roughness predicted by the

Englehund and Hansen equation is overestimated, despite the local calibration. The Colebrook and White equation gave the best results; Strickler was also very close to the field-measured values. Strickler equations may be used to estimate roughness from the grain size of the channel bed. The Englehund and Hansen equation is more suitable for sedimentation hydraulics where roughness affects the transport capacity of the flow. In the equations mentioned, the flow is assumed to be *fully developed turbulent flow*. This flow is most likely the case for natural surfaces where the roughness element heights are of the same order of magnitude as the flow depths.

Limited field tests do exist where measurements of runoff depth and velocity are used to calculate hydraulic roughness coefficients for overland flow areas. Emmett (1970) studied the hydraulics of overland flow on hillslopes at seven field sites ranging in length from 12.5 to 14.3 m. Runoff flow depth and flow velocity were measured every 0.6 m downslope. Similar measurements at 0.3-m intervals were made in laboratory test sites that had various slopes and roughness over a length of 4.6 m. From these measurements, Manning's roughness coefficients were calculated at the specified intervals. This study is one of a very few in which, at least at the hillslope scale, detailed field measurements were made to establish the hydraulic roughness. Velocity and flow depth were measured downslope under uniform artificial rainfall. Hydraulic roughness was then calculated using the Manning and Chézy equations. Figure 6.2 shows the spray from a rainfall simulator setup over the Boulder Lake Site 2 experiment reported by Emmett (1970).

Emmett's results revealed a broad range of variation of hydraulic roughness coefficients even over relatively short length scales. Table 6.1 shows the measured n-values for the Boulder Lake Site 2 hillslope, which is at a 9.6 % slope with a constant 194 mm/h artificial rainfall. In the lower half of the slope, from 3.4 to 7.6 m, the roughness averages 0.477, with a standard error of mean of 0.32, which is highly variable. The variation, as analyzed by Vieux et al. (1990) and Vieux and

Fig. 6.2 Rainfall simulator for measuring hydraulic roughness at Boulder Creek Site 2. Emmett (1970)

Table 6.1 Measured n-values for Boulder Lake Site 2

Distance (m)	Depth (mm)	Velocity (cm/s)	n-value
0.30	0.0203	3.627	0.0628
0.91	0.0660	3.353	0.1491
1.52	0.1295	2.835	0.2751
2.13	0.1041	4.938	0.1366
2.74	0.3531	1.859	0.8127
3.35	0.3124	2.591	0.5419
3.96	0.3251	2.591	0.5009
4.57	0.3658	3.018	0.5170
5.18	0.3531	3.536	0.4301
5.79	0.4521	3.078	0.5811
6.40	0.5461	2.835	0.7203
7.01	0.3277	5.151	0.2808
7.62	0.4699	3.901	0.2424

Farajalla (1994), revealed a high degree of variation in measured roughness coefficients. It may be difficult to summarize the test results, which relied on hand measurement of very shallow flow depths, because measurement errors are likely. For some experiments, the roughness was as low as 0.02 on slopes of 1.7 % and greater than 1.0 on slopes over 9 %. The extreme variability measured by Emmett (1970) differs from most published ranges, which are narrowly defined for surface types and conditions.

Modeling runoff in natural or agricultural areas must account for extreme variability of the bare and vegetated soil surfaces typically present. Surface micro-relief induced by tillage was studied by Gilley and Finkner (1991) for six selected tillage types. Random roughness parameters were used to characterize surface micro-relief. Height measurements were employed for calculating random roughness. To reduce the variation among measurements, the effects of slope and oriented tillage tool marks were mathematically removed. The upper and lower 10 % of the readings were also eliminated to minimize the effect of erratic height measurements on the final result. Random roughness (RR) measurements agreed closely with values reported in the literature. Surface runoff on upland areas was analyzed using hydraulic roughness coefficients.

Darcy-Weisbach and Manning hydraulic roughness coefficients were identified by Gilley and Finkner (1991) for each soil surface. Hydraulic roughness coefficients were obtained from measurements of discharge rate and flow velocity. The experimental data were used to derive regression relationships, which related Darcy-Weisbach and Manning hydraulic roughness coefficients to random roughness and Reynolds number.

Relating hydraulic roughness to a mapped characteristic is much more prone to subjective errors than the purely hydraulic considerations. Beyond the variation associated with land use/cover map classifications, the addition of rainfall may affect the random roughness, thus adding a temporal component. Zobeck and Onstad (1987) described a relative random roughness term (RRR) related to rainfall as:

$$RRR = 0.89e^{-0.026Rc} \qquad (6.7)$$

where Rc is cumulative rainfall in centimeters. The above equation can be used to estimate random roughness of a surface following rainfall from information on cumulative rainfall since the last tillage operation. Three criteria were established for the model equations used to predict hydraulic roughness coefficients:

1. Equations should be simple and easily solved using the fewest number of independent variables necessary to obtain reasonable results.
2. Independent variables should be generalized and applicable to conditions beyond those found in the present study.
3. Independent variables used in the relationships should be easily identified at other locations.

Variables that could significantly affect hydraulic roughness coefficients include random roughness, Reynolds number, slope, type of implement operation, and hydraulic radius. However, not all these variables would be useful as generalized predictors. In Gilley and Finkner (1991), no common basis existed for relating the six tillage implements to other machinery. Information from the six tillage treatments was used to derive the following regression equation for estimating Darcy-Weisbach hydraulic roughness coefficients:

$$f = \frac{6.30RR_0^{1.75}}{R_n^{0.661}} \qquad (6.8)$$

where RR_0 is random roughness in mm and R_n is the Reynolds number. In deriving this equation, RR_0 values varied from 6 to 32 mm (see Table 6.2), while the Reynolds number ranged from 20 to 6,000. If rainfall occurred since the last tillage operation, RR after rainfall should be substituted for RR_0 in the equation to obtain the new Darcy-Weisbach roughness coefficient.

Total hydraulic roughness on a site is usually a composite of roughness coefficients caused by several factors. Gilley and Finkner (1991) examined hydraulic roughness coefficients induced by surface micro-relief. A field study was conducted to identify random roughness and corresponding hydraulic roughness coefficients over a wide range of conditions. Random roughness measurements were made following six tillage operations performed on initially smooth soil surfaces. Random roughness measurements were found to be similar to previously reported values. Multiple linear regression analysis was used to identify the independent variables influencing hydraulic roughness coefficients. Hydraulic roughness coefficients were found to be significantly affected by random roughness and Reynolds number.

Table 6.2 Random roughness (RR$_0$) measurements (Gilley and Finkner 1991; Zobeck and Onstad 1987)

Tillage operation	Random roughness (mm)	Random roughness (mm)
Large offset disk	50	–
Moldboard plow	32	32
Lister	25	–
Chisel plow	23	21
Disk	18	16
Field cultivator	15	14
Row cultivator	15	–
Rotary tillage	15	–
Harrow	15	–
Anhydrous applicator	13	8
Rod weeder	10	–
Planter	10	6
No-till	7	–
Smooth surface	6	–

Manning hydraulic roughness coefficients may be presented as a function of Reynolds number for various tillage tools. Hydraulic roughness coefficients generally decrease with greater Reynolds number. Surfaces with the largest random roughness values usually have the greatest hydraulic roughness coefficients. The following regression equation for predicting Manning hydraulic roughness coefficients was obtained using data from six tillage treatments:

$$n = \frac{0.172 RR_0^{0.742}}{R_n^{0.282}} \tag{6.9}$$

where RR$_0$ and R_n are as previously defined. The reliability of the equation for estimating hydraulic roughness coefficients was evaluated and had a coefficient of determination, $r^2 = 0.727$.

Random roughness values available in the literature can be substituted into the regression equations to estimate Manning hydraulic roughness coefficients for a wide range of tillage implements. The accurate prediction of hydraulic roughness coefficients improves our ability to understand and properly model upland flow hydraulics in agricultural areas. Table 6.2 presents random roughness measurements where values range from 6 to 32 mm in height.

This range of roughness is presented here to illustrate that a wide range of values may exert influence on the runoff process for just one land use classification, e.g., cropland. The time since the last tillage and the amount, duration, and intensity of rainfall together with soil properties will moderate the range of hydraulic roughness present.

Table 6.3 Manning's roughness values for various field conditions

Field condition	n-value
Fallow	
Smooth, rain packed	0.01–0.03
Medium, freshly disked	0.1–0.3
Rough turn plowed	0.4–0.7
Cropped	
Grass and pasture	0.05–0.15
Clover	0.08–0.25
Small grain	0.1–0.4
Row crops	0.07–0.2

Channel roughness is usually less than overland flow and should be based on roughness elements within the channel cross-section. Barnes (1967) reported channel roughness from 50 river cross-sections measured during high flow that ranged from 0.024 to 0.075. Dingman (2009, pp. 246–249) provided guidance for stream channels and methods for estimating roughness affected by sinuosity, bed material size, vegetation, obstructions, and geometric irregularity. Because direct measurement of hydraulic roughness is difficult, especially over a watershed of any size, the derivation from maps of land use/cover is an attractive approach. In Table 6.3, Manning's n values are presented that were used in one of the first distributed hydrologic models called ANSWERS (Huggins and Monke 1966). Engman (1986) is another commonly cited source of Manning's roughness coefficients for a range of vegetative cover largely derived from studies by Ree and Crow (1977), which were developed from field hydraulic trials presented in Table 6.4.

6.3   Watershed Applications

Estimation of hydraulic roughness from land use classifications requires assigning representative values for surface runoff. Sources include local studies that may rely on satellite remote sensing, or supervised classification of aerial photography to interpret land use/cover and impervious cover. Various degrees of uncertainty exist with any of these methods, especially as it relates to impervious cover defined by land use/cover or by regression relationships tied to residential density (Sutherland 2000; Huber and Cannon 2002; Ackerman and Stein 2008). The National Land Cover Database (Homer et al. 2012, 2015) and impervious cover, which may be useful in assigning roughness values related to urban surfaces) *directly connected* to the drainage network, cause the remotely sensed imperviousness to be somewhat higher than values produced through model calibration (Huber and Cannon 2002). Estimates of a regional impervious area are dependent on method and locale, making general recommendations difficult. For example, Table 6.5 presents the range of Total Impervious Area (TIA) selected from Bochis (2007) and Bochis and Pitt (2009).

Table 6.4 Recommended Manning's coefficients for overland flow

Cover or treatment	Residue rate (ton/acre)	n-value	Range
Concrete or asphalt	–	0.011	0.01–0.013
Bare sand	–	0.01	0.010–0.016
Graveled surface	–	0.02	0.012–0.03
Bare clay—loam (eroded)	–	0.02	0.012–0.033
Fallow—no residue	–	0.05	0.006–0.16
Chisel plow	<0.25	0.07	0.006–0.17
	<0.25–1	0.18	0.07–0.34
	1–3	0.30	0.19–0.47
	>3	0.40	0.34–0.46
Disk/harrow	<0.25	0.08	0.008–0.41
	0.25—1	0.16	0.10–0.25
	1–3	0.25	0.14–0.53
	>3	0.30	–
No till	<0.25	0.04	0.03–0.07
	0.25–1	0.07	0.01–0.13
	1–3	0.30	0.16–0.47
Moldboard plow (Fall)	–	0.06	0.02–0.10
Coulter	–	0.10	0.05–0.13
Range (natural)	–	0.13	0.01–0.32
Range (clipped)	–	0.10	0.02–0.24
Grass (bluegrass sod)	–	0.45	0.39–0.63
Short grass prairie	–	0.15	0.10–0.20
Dense grass	–	0.24	0.17–0.30
Bermuda grass	–	0.41	0.30–0.46

1 ton/acre = 2,241.7 kg/ha

Table 6.5 Land use for Birmingham AL and estimates of TI, DCIA and pervious area (Bochis 2007)

Land use	TIA (%)	DCIA (%)	Pervious area (%)
High density residential	30	19	70
Medium density Residential	22	13	78
Low density residential	18	9	83
High rise res/apartments	42	17	58
Multi family	35	27	65
Commercial	73	72	27
Institutional	46	41	54
Industrial	59	50	41
Open space	13	9	87
Freeways	58	0	42

The directly connected impervious area is less than the TIA and follows a power law relationship to the TIA but again, varies considerably from region to region.

An implicit assumption of reclassification of NLCD classes is that the classification represents the vegetation that runoff encounters as it travels overland. This assumption may not be valid since the classification for forest does not define the understory or how dense any litter may be on the ground surface. Assignment of hydraulic roughness based on NLCD land use classification codes is dependent on field knowledge of the watershed and locale. Kalyanapu et al. (2010) tested watershed model prediction for a 23 km^2 watershed near Houston TX in Greens Bayou. Performance was compared at the outlet and at an internal subcatchment using two sources; (1) visual inspection and (2) NLCD roughness assignment. Minimal differences were found at the outlet, whereas divergence was found in subcatchments. The main cause of this internal divergence was thought to be that the NLCD classes did not discriminate adequately especially where there was residential or commercial development. Table 6.6 presents their LULC and hydraulic roughness assignment based on the McCuen (1998) hydraulic roughness by land use class. Kalyanapu et al. (2010) assigned areal weighted average Manning's n values for Dense Grass/Light Woods as $n = 0.2$; Short Grass/Lawn, $n = 0.15$; and asphalt and concrete surfaces as $n = 0.012$ and 0.013.

As with infiltration parameters, perhaps the most important aspect is to assign initial estimates that are representative of the relative roughness values (high/low). Table 6.7 presents the assumed n-values and the basis for assignment to each class with source notes. Neither McCuen (1998) nor the references cited therein offer suggestions for forest land cover, so an interpretation is to assume $n = 0.10$ (Range Land, clipped) as representative of the understory (as in Fig. 6.1). This is less rough compared to the McCuen value used by Kalyanapu et al. (2010), perhaps

Table 6.6 Watershed scale hydraulic roughness based on NLCD classification (Kalyanapu et al. 2010)

Land cover	Description	Manning's n
21	Developed, open space	0.0404
22	Developed, low intensity	0.0678
23	Developed, medium intensity	0.0678
24	Developed, high intensity	0.0404
31	Barren land	0.0113
41	Deciduous forest	0.36
42	Evergreen forest	0.32
43	Mixed forest	0.40
52	Shrub/scrub	0.40
71	Grassland/herbaceous	0.368
81	Pasture/hay	0.325
90	Woody wetlands	0.086
95	Emergent herbaceous wetlands	0.1825

Table 6.7 NLCD classification with suggested n-value and source notes

LC	Area (%)	Count	Class description	n-value	Source notes
11	3.0	1226	Open water	0.000	No resistance interpreted for water
21	3.9	1607	Developed open space—Impervious <20 %	0.162	Crawford and Linsley (1966)—light turf/concrete surface
22	1.0	434	Developed, low intensity—Impervious 20–49 %	0.134	Crawford and Linsley (1966)—light turf/concrete surface
23	0.3	123	Developed, medium intensity—Impervious 50–79 %	0.078	Crawford and Linsley (1966)—light turf/concrete surface
24	0.1	44	Developed, high intensity—Impervious 80–100 %	0.031	Crawford and Linsley (1966)—light turf/concrete surface
31	0.2	78	Barren land (Rock/Sand/Clay)—Vegetation <15 %	0.045	Engman (1986)—Bare clay-loam
41	50.7	21055	Deciduous forest	0.100	Engman (1986)—Range (clipped)
42	0.7	301	Evergreen forest	0.100	Engman (1986)—Range (clipped)
43	1.0	405	Mixed forest	0.100	Engman (1986)—Range (clipped)
52	0.8	340	Shrub/scrub	0.100	Engman (1986)—Range (clipped)
71	1.1	474	Grassland/herbaceous	0.100	Engman (1986)—Range (clipped)
81	20.7	8600	Pasture/hay	0.100	Engman (1986)—Bermuda
82	0.6	240	Cultivated crops	0.080	Engman (1986)—Disk Harrow
90	0.6	251	Woody wetlands—Forest or shrub land vegetation	0.400	Crawford and Linsley (1966)—Dense shrubbery and forest litter
95	0.0	13	Emergent herbaceous wetlands	0.400	Crawford and Linsley (1966)—Dense shrubbery and forest litter
			Basin average	0.086	

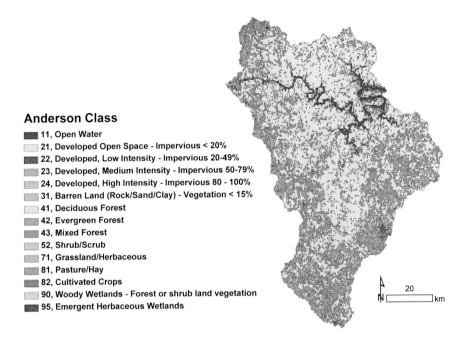

Anderson Class
- ■ 11, Open Water
- ▨ 21, Developed Open Space - Impervious < 20%
- ■ 22, Developed, Low Intensity - Impervious 20-49%
- ▨ 23, Developed, Medium Intensity - Impervious 50-79%
- ▨ 24, Developed, High Intensity - Impervious 80 - 100%
- ▨ 31, Barren Land (Rock/Sand/Clay) - Vegetation < 15%
- ▨ 41, Deciduous Forest
- ▨ 42, Evergreen Forest
- ▨ 43, Mixed Forest
- ▨ 52, Shrub/Scrub
- ▨ 71, Grassland/Herbaceous
- ▨ 81, Pasture/Hay
- ■ 82, Cultivated Crops
- ▨ 90, Woody Wetlands - Forest or shrub land vegetation
- ■ 95, Emergent Herbaceous Wetlands

Fig. 6.3 NLCD classification for the Osage River and Niangua tributary in central Missouri

attributable to more densely vegetated coastal watersheds near Houston TX. Figure 6.3 shows the 2006 NLCD for a 6,366 km^2 basin located in central Missouri for a reach of the Osage River. The dominant classes are Forest (Classes 41, 42, 43) and Pasture/Hay (Class 81) comprising 89 % of the basin. These classes control the basin average. After model calibration to observed streamflow, the initial estimate of overland roughness was revised from $n = 0.086$ to $n = 0.099$, which is a small change due to the selected initial value of $n = 0.10$ for the main class, Forest and Pasture/Hay in Table 6.7.

6.4   Summary

Hydraulic roughness parameter maps are a key factor in controlling the velocity at which runoff travels through the drainage network and downstream reaches of the watershed. Direct measurement is usually not practical at watershed scales. Hydraulic roughness can be inferred from land use/cover. Assigning hydraulic roughness to general land use/cover classifications is difficult, because these classifications are rarely made according to hydrologic characteristics. The hydrologist must exercise some judgment in assigning hydraulic roughness parameters to these classification schemes and often requires interpretation of what a particular land use/cover classification may indicate in terms of overland flow roughness. Field observation or experience is usually required to assign appropriate values of hydraulic roughness to channels and overland flow areas. While the values shown in a map may be used to simulate runoff in the overland flow areas of the basin, channel hydraulic roughness must be supplied from other sources. Channel roughness should be based on representative channel conditions, rather than land use/cover classifications. Impervious areas mapped by land use/cover maps can be used for assigning hydraulic roughness to such areas, besides the obvious implications for infiltration modeling. Initial values are used for distributed model setup and then adjusted by calibration. When developing these initial parameter maps, assigning relatively *smooth* versus *rough* parameter values is more important than the accuracy of the values assigned.

References

Ackerman, D. and Stein, E.D. 2008. Estimating the Variability and Confidence of Land Use and Imperviousness Relationships at a Regional Scale. *JAWRA Journal of the American Water Resources Association*, 44: 996–1008. doi: 10.1111/j.1752-1688.2008.00215.x

Barnes, H.H. 1967. *Roughness characteristics of natural channels* (No. 1849). US Govt. Print. Off.

Bochis, E.C. 2007. *Magnitude of impervious surfaces in urban areas.* (Doctoral dissertation, University of Alabama).

Bochis, C. and R.E. Pitt. 2009. January. Land use and runoff uncertainty. In *World Environmental and Water Resources Congress 2009@ Great Rivers*, 1314–1324. ASCE.

Chow, V.T., D.R. Maidment, and L.W. Mays. 1988. *Applied hydrology.* New York: McGraw-Hill.

Crawford, N.H. and R.K. Linsley. 1966. Digital Simulation in Hydrology: Stanford Watershed Model 4.

Dingman, S.L. 2009. *Fluvial hydraulics*, p. 559. New York: Oxford University Press.

Emmett, W.W. 1970. *The hydraulics of overland flow on hillslopes.* U.S. Geological Survey Professional Paper No. 662-A U.S. Washington D.C: Govt. Printing Office.

Engelund, F. and E. Hansen. 1967. *A monograph on sediment transport in alluvial streams.* TEKNISKFORLAG Skelbrekgade 4 Copenhagen V, Denmark.

Engman, E.T. 1986. Roughness coefficients for routing surface runoff. *Journal of Irrigation and Drainage Engineering* 112(1): 39–53.

Gilley, J.E., and S.C. Finkner. 1991. Hydraulic roughness coefficients as affected by random roughness. *Soil and Water Division of ASA* 34(3): 897–903.

Homer, C.H., J.A. Fry, and C.A. Barnes. 2012. The national land cover database. *US Geological Survey Fact Sheet* 3020(4): 1–4.

Homer, C.G., J.A. Dewitz, L. Yang, S. Jin, P. Danielson, G. Xian, J. Coulston, N.D. Herold, J.D. Wickham, and K. Megown. 2015. Completion of the 2011 National Land Cover Database for the conterminous United States-Representing a decade of land cover change information. *Photogrammetric Engineering and Remote Sensing* 81(5): 345–354.

Huber, W.C. and L. Cannon. 2002. Modeling non-directly connected impervious areas in dense neighborhoods. *Global Solutions for Urban Drainage* 8–13.

Huggins, L.F. and Monke, E.J., 1966. *The mathematical simulation of the hydrology of small watersheds.* Technical Report No. 1, 130 pp. West Lafayette: Purdue University Water Resources Research Center, Indiana.

Kalyanapu, A.J., S.J. Burian, and T.N. McPherson. 2010. Effect of land use-based surface roughness on hydrologic model output. *Journal of Spatial Hydrology* 9(2).

McCuen, R.H. 1998. *Hydrologic analysis and design.* New Jersey: Prentice-Hall.

Ree, W.O., and F.R. Crow. 1977. *Friction factors for vegetated waterways of small slope.* US Department of Agriculture: Agricultural Research Service.

Sutherland, R. 2000. Methods for estimating the effective impervious area of urban watersheds. *The practice of watershed protection*, 193–195. Center for Watershed Protection.

Vieux, B.E., V.F. Bralts, L.J. Segerlind, and R.B. Wallace. 1990. Finite element watershed modeling: one dimensional elements. *Journal of the Water Resources Planning and Management* 116(6): 803–819.

Vieux, B.E., and N.S. Farajalla. 1994. Capturing the essential spatial variability in distributed hydrological modeling: Hydraulic roughness. *Journal of Hydrological Processes* 8: 221–236.

Zobeck, T.M., and C.A. Onstad. 1987. Tillage and rainfall effects on random roughness: A review. *Soil and Tillage Res.* 1: 1–20.

Chapter 7
Watersheds and Drainage Networks

Abstract Grid-based distributed hydrologic models rely on a drainage network to model basin response. A digital elevation model (DEM) is useful for characterizing the terrain and drainage network. Processing steps and DEM cell size affect the land surface slope, drainage network length and connectivity properties. Derived drainage networks and the hydraulic parameters used to represent the conveyance of runoff to the outlet of the river basin are dependent on cell size and on the methods used to derive the drainage network. Once the drainage network is defined and slope derived, the remaining hydraulic parameters are applied to overland and channel cells, which may require adjustment. The characteristics of the extracted drainage network can influence hydrologic model calibration and performance.

7.1 Introduction

Distributed hydrologic models that simulate rainfall-runoff require some type of data structure, called a drainage network to route runoff through the topography. The drainage network in Fig. 7.1 shows the connectivity derived from a DEM by connecting each grid cell together according to the principal direction of slope. The highlighted elements on the right correspond to the drainage area contributing to a gauging station located along the Kee Lung River that runs through Taipei, Taiwan (Vieux et al. 2003). A drainage network is composed of channel and overland flow elements that represent flow through and over the terrain. These elements are typically represented as grid cells in a raster data structure.

Whether the DEM is gridded or composed of a triangular irregular network (TIN), DEM resolution is important. Further, as a distributed model is calibrated at one resolution, the calibrated parameters may require adjustment as larger grid-cell sizes are used. Automatic extraction of drainage networks from DEMs must consider whether a cell is a channel or overland flow cell. Model performance and calibration can be affected by assumptions used to extract the network, especially when selecting the channel cells as a percentage, e.g., 10 %. Seemingly arbitrary choices can produce marked changes in the hydrologic response. A dense channel

B.E. Vieux, *Distributed Hydrologic Modeling Using GIS*,
Water Science and Technology Library 74, DOI 10.1007/978-94-024-0930-7_7

Fig. 7.1 Drainage network
derived from 250-m
resolution digital elevation
model

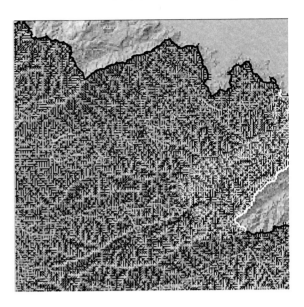

network will generally produce higher and earlier peaks in the discharge hydrograph. The following sections describe methodological approaches to delineation and the influence of resolution and processing steps on the automatic extraction of a drainage network from DEMs.

7.2   Drainage Network Extraction

Accurate representation of the drainage network that connects hillslopes to channels and then to the basin outlet must capture the spatially distributed topographic information besides other factors affecting the hydrologic process. A raster DEM contains topographic information as a regular array of elevation data. Although other data structures such as TINs are in use, a vast amount of raster data is available in the US and worldwide at various resolutions. Spatial variability of the topography represented by the DEM is affected by the source of the data, the original map scale and the resolution at which the data are compiled. Assuming that the numerical algorithm employed in the model is stable and accurate, the focus of deterministic, distributed modeling should be on selecting the size of the computational element (grid cell or otherwise) that accurately represents the natural features affecting the process.

Distributed modeling of hydrologic processes relies on discrete representation of both continuous (e.g., topography) and discontinuous (e.g., land use/cover) surfaces. The drainage network consisting in both channel and overland flow segments is derived by connecting each grid cell in the DEM. The length of this derived drainage network scales according to a fractal scaling law. The drainage network controls the rate at which runoff is routed to the outlet. Thus, grid-cell resolution profoundly affects topographically based distributed hydrologic models.

Delineation of drainage networks for distributed hydrologic modeling using DEMs was reported by Moore and Grayson (1991), Quinn et al. (1991), Tarboton et al. (1991), Chang and Tsai (1991), Vieux (1993, 1988), Vieux and Needham (1993). Water quantity and quality modeling described by these authors illustrate the dependence of the simulation on the quality of the delineated watersheds and drainage network. Other difficulties arise from model assumptions and structures that are dependent on such arbitrary choices as the size of the subbasins and routing. Quinn et al. (1991) described the development, applications, and limitations of TOPMODEL used with drainage pathways derived from a raster DEM. They found that model sensitivity to the grid-cell size yielded inaccurate results as cell size increased, particularly on divergent hillslopes. Tarboton et al. (1991) discussed the importance of map scale in the delineation, validation, and use of the drainage networks derived from DEMs. Chang and Tsai (1991) investigated the effect of spatial resolution of the DEM on the derived slope and aspect maps. Vieux (1993) found that the apparent drainage length and slope of a watershed decreased with increased grid-cell size, propagating error in hydrograph simulations. Vieux (1993), Vieux and Farajalla (1994), Farajalla and Vieux (1995) investigated grid-cell size effects on the spatial variability of topography, hydraulic roughness, and infiltration parameters, respectively. These studies described the development and application of a method for assessing the grid-cell size that captures the variability of hydrologic parameters (cf. Chap. 4).

Jenson and Dominique (1988), Jenson (1991), Martz and Garbrecht (1992), Freeman (1991), Hutchinson (1989) discussed the automatic delineation of drainage directions from DEMs and problems associated with the delineation process. Jenson and Dominique (1988) developed a drainage analysis technique that helps in delineating basins in flat terrain where flow direction is ambiguous due to pits, sinks, and/or dams. Freeman (1991) proposed a basin delineation method that incorporates divergent flow into delineation of flow paths. Hutchinson (1989) developed a morphological approach to digital elevation modeling and catchment delineation that is advantageous for use in hydrologic modeling.

The basic steps in extracting hydrologic features from a DEM involve the following:

1. Depressions are filled
2. Flow direction in four or eight directions is computed based on principal direction of flow
3. Flow accumulation is computed for each cell
4. Slope is computed along the principal direction of slope
5. Stream channels are assigned based on the flow accumulation
6. Watershed boundaries are delineated that encompass the stream network.

Extracting drainage networks and other hydrologic features or characteristics, such as drainage accumulation and slope, are essential for distributed hydrologic modeling at the basin scale.

Lee and Chu (1996) pointed out that the extraction of hydrological features from a DEM has become the de facto procedure in many GISs due to two recent trends: GIS functions for processing DEM data are becoming easier to use and DEM data are becoming increasingly available. Processing DEM data to extract hydrological features has become a routine operation. While having worldwide geospatial data is desirable, coarse resolution DEMs of low quality may lead to unreliable analytic results. It is known that DEMs contain errors but it is not generally known how these errors impact the results derived from using DEM data.

Analysis of the impact of potential errors on the extracted hydrological features from DEMs was performed using a simple simulation study (Lee and Chu 1996). For the test data, the authors selected a set of US Geological Survey (USGS) DEMs based on their spatial structures, at the scale of 1:250,000. They used a set of computerized procedures for extracting drainage cells as the basis for comparative study. In addition, random errors of various magnitudes were simulated and added to these DEMs. Similar routines were then applied to extract drainage cells from these simulated DEMs. The results were analyzed to reveal the effect of simulated errors and the spatial structure of DEMs on extracting drainage cells. Spatial autocorrelation is a measure of how close similar values cluster together. Each element in a DEM carries a measure of the terrain, i.e., elevation at the location of that cell. A DEM is said to have a high level of spatial autocorrelation if neighboring cells display similar measures of the terrain. Conversely, a more rugged surface would display a lower degree of spatial autocorrelation.

For each DEM examined by Lee and Chu (1996), elevation errors were found to affect the extracted drainage cells. The impact of uncertainty in elevation depended on the degree of spatial autocorrelation. The influence was the most severe for DEMs of higher spatial autocorrelation. Extracting hydrological features using DEM data is very sensitive to the potential errors in digital terrain elevation data. A slight distortion of the terrain measures can lead to dramatic differences in the resulting hydrological features. Careful consideration should be given to features extracted from DEM data.

Watershed delineation and associated drainage networks may be identified from a DEM using algorithms that use grid-cell elevations to find flow directions. Extracting a drainage network from a DEM requires some definitions for the algorithm to proceed effectively. Fern et al. (1998) reported problems with the detection of false stream segments and attributed these errors to the low resolution of the level-1 DEM used for testing. The network performance is hoped to improve as the resolution of the data is increased. The main advantages of this method over previous local operator methods are that it allows the extraction of lakes as well as stream segments and it is not dependent on the resolution of the DEM (assuming sufficient sampling) or the width of stream segments. Fern et al. (1998) defined the following:

1. A point must be a member of a valid valley segment (river or stream), or a larger drainage basin (lake or pond).
2. Elevations along a valley segment must decrease in one direction (flow direction) since water flows from higher to lower elevations.
3. A valid valley segment must have a source where water can enter into and travel in the flow direction of the particular segment. A source may originate at the junction of DEM cells, another valley segment, or a point at the end of the segment whose elevation is the highest in the segment.
4. A valid valley segment usually has a decision for its flow of water. In other words, the water flow cannot stop abruptly. A destination can be another valley segment, a larger drainage basin, or the junction of other DEM cells.

Implementation of an extraction algorithm is complicated by errors in the DEM that make finding a connected drainage network difficult. How DEM errors affect feature extraction algorithms, or how they are overcome, may be found in Lee et al. (1992), Fern et al. (1998), and in Wilson (2012).

Most extraction algorithms apply local operators to a 3 × 3 kernel of cells. Many of these techniques suffer from deficiencies that are a result of the local nature of the algorithms. Perhaps the most familiar algorithm for drainage network extraction is the one described by O'Callaghan and Mark (1984). Beginning with assignment of drainage direction to each pixel, an iterative computation is then performed in which drainage accumulation values are updated for each pixel, based on a weighted sum of the accumulation values of surrounding pixels. Jenson (1985) used a moving 3 × 3 pixel operator to label possible drainage points by searching for local minima between nonadjacent pixels. Localized rules are then used to extract a possible drainage network, based on a user-specified distance and elevation threshold. The extracted network that results is often broken into unconnected fragments making it difficult to achieve the global reasoning necessary to establish links between separated stream segments. Also, these algorithms may not be able to distinguish local minima that are actual topographic features such as pools or depressions in the DEM that are not part of the overall drainage network. Finally, the localized operator methods are dependent on the DEM resolution and the widths of the drainage features. The resolution deficiency can be partly overcome by down sampling the DEM when necessary. The width deficiency, however, cannot be dealt with easily, since drainage features across the DEM cell are not of uniform width. This means that down sampling a DEM cell by a certain factor aids in the extraction of some drainage features while hurting the extraction of others.

An expert system-based method that uses both local operators and global reasoning to solve for a valid drainage network was described by Qian et al. (1990). First, using a local operator and a reasoning process, groups of pixels are labeled as possible stream segments and are given to a hypothesis generator. The hypothesis generator suggests links between spatially related segments and decides which segments are not parts of the overall drainage network. This more global approach to drainage network extraction produces results far superior to the previously described local algorithms. This approach still uses a local operator to make an initial guess at the drainage network.

Garbrecht and Shen (1988) reviewed the physical basis of the linkage between magnitude and timing of channel flow hydrographs and drainage network morphometry. The physical framework of the dependence between channel flow hydrographs and drainage network morphometry surface runoff takes place on subcatchments connected by channels to route flow downstream. Size, shape, slope, number, and spatial arrangement of individual subcatchments and channels control the collection, storage, routing, and concentration of rainfall-runoff into a channel flow hydrograph. The geometric constraints and the topologic properties of the drainage network are the roots of the linkage between the channel flow hydrograph characteristics and the network. For the Hortonian networks the variability of the geometry of individual channels and subcatchments within each Strahler order was generally found to have little effect upon the overall character of the hydrograph in channels of higher order. In their approach, the formation of runoff, traveltime, and concentration of the hydrographs can be simplified to a sequence of the representative hydrograph based on channel order.

The analysis of large river networks using digital elevation models has given insight into the variation of channel slopes with scale. Investigators recently suggested that channel slopes are self-similar with magnitude or area as a scaling parameter. Tarboton et al. (1989) found that the variance of channel slope was larger than predicted by simple self-similarity, indicating multi-scaling fractal dimension. The scaling exponent for the standard deviation is approximately half the corresponding exponent in the relationship between mean slope and the drainage area. The delineation of watersheds from a DEM was described by Martz and Garbrecht (1992); multi-scaling of delineated drainage networks by Tarboton et al. (1989); the algebra for terrain-based flow analysis in Tarboton and Baker (2008); and parallel processing procedures in Tesfa et al. (2011).

7.2.1 Sensitivity to Drainage Network Composition

Distributed models rely on the drainage network composed of overland and channel elements. The division between overland and channel cells has both practical and theoretical importance. Automatic algorithms for extraction of drainage networks must assign grid cells as channels or overland flow. The portion of the drainage network comprised of channel cells is termed the *channel network*. Model response is affected by the number of grid cells that are considered channel cells as opposed to overland flow. If a large proportion of the drainage network is designated as a channel, then hydraulic characteristics must be input for each channel reach or even every cell. Practically, using a large proportion of channel cells may not be feasible unless many channel cross-sections are available or the hydraulic characteristics are known. Without detailed channel hydraulic measurements, a relationship can be derived that relates channel hydraulic properties to drainage area within geomorphic regions. In any case, the theoretical importance of increasing the proportion of channel cells compared to overland flow cells is fundamental to the adequate representation of the hydraulics affecting the physics-based hydrologic response. Fewer channels in proportion to overland cells tend to cause hydrographs with delayed and attenuated peak discharge.

An effect of constructing channels is the acceleration of runoff to the outlet of a basin, which is caused by the improved hydraulic efficiency of a channel compared to overland flow. So too in a model, if overland flow cells are replaced by channel cells, we should expect that the hydrologic response of the modeled basin will change. Channelization causes the hydrologic response to behave differently because the hydraulics affecting the routing of runoff is changed. Thus, when more channels are included relative to overland flow cells, the model will respond with higher peaks and earlier time-to-peaks. Figure 7.2 shows a drainage area composed only of overland flow elements, whereas Fig. 7.3 has been channelized with the addition of channel elements along the major drainage paths in this hypothetical watershed.

The percentage of channel versus overland cells is usually controlled by assigning a flow accumulation threshold, or assigned because of mapped stream location. Once the stream channels are identified, then hydraulic properties must be assigned that are representative of the channels in the watershed. To illustrate the influence of channel cells, a progressive replacement of overland cells with channel cells is performed as shown in Fig. 7.3.

The simulated response from this hypothetical watershed with progressively more channelization results in the hydrographs shown in Fig. 7.4. The differences between the hydrographs are due solely to the degree of channelization. The highest

Fig. 7.2 Drainage network composed only of overland flow elements (no channel elements were added)

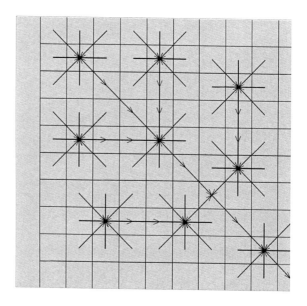

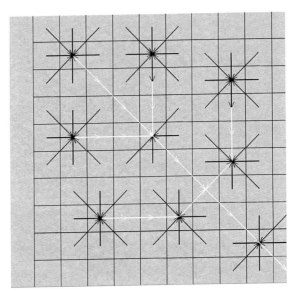

Fig. 7.3 Drainage network composed of channel and overland flow elements. Channel elements are shown as *white elements* with the outlet cell in the *lower right corner*

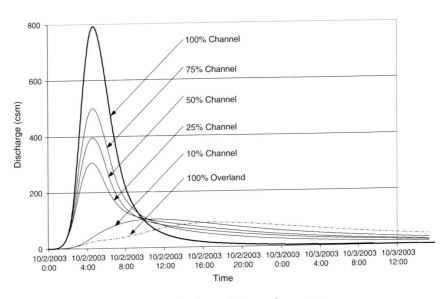

Fig. 7.4 Effect of progressive channelization on hydrograph response

response (dark line) is from the channelized network, whereas the hydrograph shown by the dashed line is from the watershed represented by all overland flow elements. The percentage of channel versus overland cells can become an implicit parameter affecting model response. As the percentage of channel cells is increased, the drainage network becomes more efficient at routing runoff to the outlet compared to overland flow cells as evidenced by the increased peak discharge values.

7.2.2 *Resolution-Dependent Effects*

A drainage network is the network formed by connecting each cell according to the steepest descent to one of the eight nearest neighboring cells. The total length is the sum of overland and channel lengths. In other words, it is the length traveled by the water in the basin before reaching the basin outlet. As grid-cell size is increased two slope effects result: flattening due to 'cutting' the hills and 'filling' the valleys and flattening due to lengthening of the distance between two adjacent cells. Using larger DEM cell sizes also 'short-circuits' streams or river meanders, causing an overall shortening of the drainage network. Drainage length shortening and slope flattening often have competing effects on distributed simulations of hydrographs (cf. Vieux 1993). The elemental length and the number of elements used to represent a flow network directly impacts discharge values simulated using the drainage network to route runoff. Because of this dependence, the resolution of the DEM and its impact on hydrologic simulation with grid-based distributed models cannot be overlooked.

A useful framework for understanding how resolution affects extracted hydrologic features such as drainage length is found in the fractal scaling law. A *fractal* may be defined as a geometric set consisting in points, lines, areas, or volumes whose measure (e.g., length) changes with the resolution of the measurement (Goodchild and Mark 1987). If an irregular line (e.g., a coastline) is measured at two different scales, the length in almost all cases increases by more than the ratio of the scales. As new detail in the coastline becomes apparent at the larger scale, the length is increased. The fractal dimension, D, of this irregular line is given by Goodchild and Mark (1987) as:

$$D = \frac{\log(n_2/n_1)}{\log(s_1/s_2)}, \tag{7.1}$$

where n_1 and n_2 are the number of divider steps of size s_1 and s_2, respectively. Goodchild (1980) discussed a method for determining the fractal dimension of a line by measuring the length using a range of measurement intervals. The length of the measured line is plotted against map scale on log–log scale. The slope of the best-fit line is 1D, the fractal dimension. This method is commonly referred to as the *divider method*. Others have devised methods to assess the fractal dimension of natural surfaces and, in particular, river and drainage networks. Robert and Roy (1990), Hjelmfelt (1988), Tarboton et al. (1988), La Barbera and Rosso (1989), Goodchild (1980), Huang and Turcotte (1989) discussed applications of the fractal dimension in the field of hydrology. Robert and Roy (1990) found that the value of the fractal dimension for 23 drainage basins of the Eaton River in Quebec, Canada varied between 1.08 and 1.3, depending on the map scales. La Barbera and Rosso (1989) found that the geometric pattern of a stream network could be viewed as a fractal object. They found that the fractal dimension of the river networks varied between 1.5 and 2.0, with an average value of 1.6–1.7. For eight rivers in Missouri, Hjelmfelt (1988) found that the fractal dimensions varied between 1.04 and 1.45. Tarboton et al. (1988) argued that the fractal dimension of the rivers is not significantly less than 2 but this depends on the percentage of cells classified as channels.

From these applications of scaling theory, it is evident that the fractal dimension depends on the definition of the drainage network and the detail apparent in the map from which measurements are taken. A drainage network comprising both channel and overland flow segments drain the entire basin, suggesting that the fractal dimension of the river networks should approach 2 since it is entirely space filling. Disagreement on whether the fractal dimension of a river network is closer to 1.0 than 2.0 is likely due to the definition of the network. A network that connects every grid cell will have a dimension equal to 2.0 (space filling), whereas, a network showing only a few major river channels will have a dimension closer to 1. The importance of the fractal dimension is the theoretical basis for understanding why drainage length changes with resolution. If the length of the drainage network becomes either longer or shorter in absolute terms, then the hydrologic model based on the network will perform accordingly.

Slope and drainage length derived from a DEM cannot be entirely separated in practice. The impact of grid-cell resolution on slope is addressed in Chap. 4. Here we examine how the DEM resolution affects the drainage length according to the fractal scaling law represented by Eq. 7.1. To illustrate the dependence of drainage length on DEM cell resolution, a data set is derived from the 2400 km^2 Illinois River. The following steps determine the length of the drainage network:

1. Identify the D8 drainage direction for each channel and overland flow cell in the basin.
2. Compute the drainage length as the product of resolution (or cell hypotenuse) and cell count for both channel and overland cells.
3. Repeat drainage length computation for a range of resolutions.

Results of applying these steps to the Illinois River Basin are presented in Table 7.1.

The columns contain the cell resolution, the number of cells at that resolution, the drainage length connecting each cell and the fractal dimension computed by

Table 7.1 Drainage lengths at various grid-cell resolutions for the Illinois River Basin

Size (m)	Length (km)	Number	Dimension
30	132,611	3,644,838	–
60	69,665	963,887	1.9
240	18,129	62,048	2.0
480	9076	15,482	2.0
960	4483	3819	2.0

Fig. 7.5 Comparison of the
drainage lengths at different
resolutions

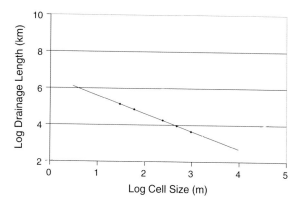

Fig. 7.6 Fractal scaling
between resolution and the
number of cells in a drainage
network

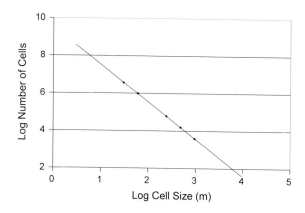

Eq. 7.1. As the resolution is increased from 30 to 960 m, the drainage length is shortened dramatically from 132,611 km to only 4,483 km for the same drainage area. Figure 7.5 shows the linear relationship on a logarithmic scale between length and cell size.

The network connects every grid cell in the basin. Plotting the logarithm of the number of cells, n and the resolution, s, reveals a linear relationship with constant slope.

Because the drainage length follows a fractal scaling law, the length varies with grid-cell resolution. The slope of the line (2.0) and its linearity is not surprising if we consider that the drainage network is space filling. Figure 7.6 shows that the number of cells follows a fractal scaling law with a constant slope, which in this example is equal to 2.0. A drainage network derived at a particular DEM resolution will have properties that can affect hydrologic simulation and model calibration. The DEM resolution and derivative slope and drainage network maps form the basis for grid-based distributed models.

7.2.3 Constraining Drainage Direction

Most automated extraction algorithms are efficient at finding a consistent and accurate drainage network using the elevation data contained in a DEM. To extract drainage features, preprocessing steps are often necessary that include filling pits and possibly smoothing the DEM before the extraction algorithm is applied. While these steps make extraction more efficient, smoothing and filling pits can cause erroneous drainage patterns in terrain with depressions. However, anomalies in the extracted drainage network do occur where coarse resolution DEMs do not accurately resolve drainage direction. Flat areas and low data precision elevation data also contribute to erroneous drainage directions and resulting networks. Another weakness stems from nonrepresentative drainage directions that are assigned based on regular sampling (grid cells) of an irregular surface (terrain). Flat areas where two or more cells have the same elevation can cause the drainage network to 'capture' a river, giving the appearance that flow is in a different direction than is actually the case.

Constraining the automatically delineated drainage network from raster DEMs can be achieved by using stream and watershed maps in vector format. By combining the raster DEM with vector hydrographic information, an improved drainage network can result that is more representative of the actual network. This process is commonly known as 'burning in' the stream network. The burn-in procedure involves artificially lowering of the elevations wherever the stream location is mapped. The watershed boundary can also be raised to produce a wall beyond which the drainage network is not permitted to extend. The result of this procedure is an extracted drainage network that follows the mapped stream network. It is also constrained by the watershed boundary. A vector stream map combined with a raster DEM can result in an improved drainage direction map. Figure 7.7 shows a vector stream map overlaid on to a raster DEM for a 5,835 km^2 basin in Romania.

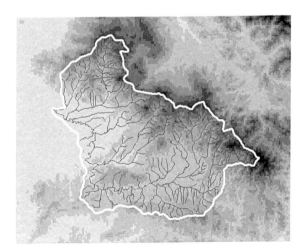

Fig. 7.7 Lapus Basin showing a 1-km resolution DEM and vector stream map overlay

Fig. 7.8 Misalignment between stream location indicated by raster flow accumulation and a vector map overlay (Courtesy, INMH, Romania)

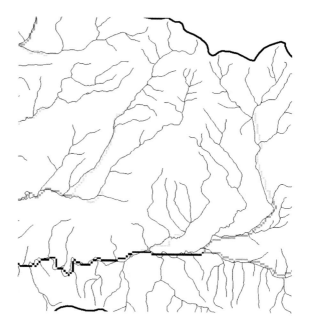

The quality of extracted drainage networks and watershed boundaries is highly dependent on the quality of the stream map used to constrain the flow direction in the extraction process. If a vector-format stream map is compiled at a different scale than the DEM, the location of the stream defined by the vector stream map may not correspond exactly with the stream defined by DEM. Because the two data sets are compiled at different scales, slight (or even major) misalignment may result between the streams defined by the raster DEM and the vector stream map. This effect can be seen in Fig. 7.8, which is zoomed into the southwest portion of the Lapus Basin. Using the vector stream map to constrain flow direction extracted from the DEM will remove this misalignment. The misalignment shown in Fig. 7.8 is not severe because it is located in mountainous terrain that is well defined at the DEM resolution.

More severe errors in extraction of hydrologic features can result in flat areas. Under such terrain conditions, adjacent cells may have the same elevation making flow direction identification difficult or erroneous. Figure 7.9 shows a drainage network derived from a 1-km resolution DEM within the Tanshui River near Taipei, Taiwan. The overland flow directions from cell to cell are shown in gray with the stream channel shown in black. The stream channel that crosses diagonally from upper right to lower left is contrary to actual flow direction. Figure 7.10 shows the resulting constrained drainage network extracted by combining a vector stream map with the DEM. The coarseness of the horizontal resolution (1 km) and the precision of the elevation (nearest 1 m) contribute to erroneous drainage network extraction in this flat area. Improved drainage network extraction can be achieved using a stream map that is known to be accurate or is consistent with the scale at

Fig. 7.9 Unconstrained
drainage direction map with
the delineated stream shown
in *black*

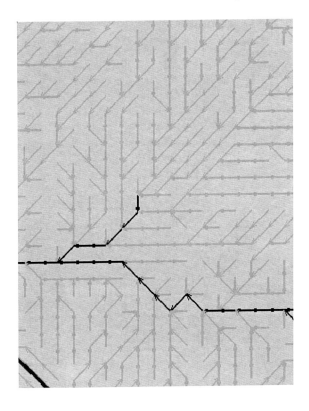

which the DEM is compiled. Higher resolution DEMs may produce reliable drainage networks even without burn-in. Once the drainage network is extracted using combined information, the channels will correspond to the vector stream map. These channel locations may require inspection to judge if the extracted network is sufficiently accurate for the desired hydrologic modeling scale.

7.3 Summary

Grid-based distributed models depend on digital elevation data to route surface runoff from cell to cell and ultimately to the basin outlet. Extracted hydrographic features depend on the grid-cell resolution of the DEM and terrain characteristics. The proportion of assumed channel cells relative to overland flow cells dramatically affects the modeled hydrograph. Model response will be affected depending on how many grid cells are considered channels. The fractal scaling provides a useful framework for understanding the dependence of drainage length on grid-cell

Fig. 7.10 Constrained
drainage direction map with
the delineated stream shown
in *black*

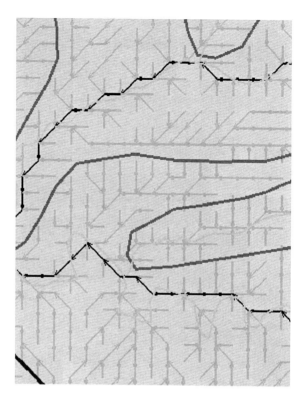

resolution. The slope and length of the extracted drainage network are major
determinants of modeled hydrologic response. Constraining the drainage network
delineation by 'burning in' a stream map into the DEM can improve the resulting
drainage network. However, care must be taken when the stream map is compiled at
a different scale than the DEM.

The problems that occur during drainage network extraction may have significant influence on hydrologic model calibration and performance. Once the drainage
network is defined and slope derived, the remaining parameters are adjusted.
However, the two steps are linked because the slope parameter assigned to the
drainage network will in turn influence the hydraulic roughness values obtained
through calibration. For example, if a coarse resolution DEM is used, then shortened drainage length and attenuated slope values will likely result. Depending on
the interrelationship of these resolution-dependent parameters, runoff may be
accelerated by the shorter drainage length or slowed by the flatter slopes. Because
the drainage network extracted from a DEM is dependent on resolution and
percentage of channel cells versus overland, hydrologic model performance and
calibration are sensitive to the resulting network of channel and overland cells.

References

Chang, K., and B. Tsai. 1991. The effect of DEM resolution on slope and aspect mapping. *Cartography and Geographic Information Systems* 18(1): 69–77.

Farajalla, N.S., and B.E. Vieux. 1995. Capturing the essential spatial variability in distributed hydrologic modeling: Infiltration parameters. *Journal of Hydrological Processes* 8(1): 55–68.

Fern, A., M.T. Musavi, and J. Miranda. 1998. Automatic extraction of drainage network from digital terrain al network approach. *IEEE Transactions on Geoscience and Remote Sensing* 36(3): 1007–1015.

Freeman, T.G. 1991. Calculating catchment area with divergent flow based on a regular grid. *Computers & Geosciences* 17(3): 413–422.

Garbrecht, J., and H.W. Shen. 1988. The physical framework of the dependence between channel flow hydrographs and drainage network morphometry. *Journal of Hydrological Processes* 2(337): 355.

Goodchild, M.F. 1980. Fractals and the Accuracy of Geographical Measures. *Mathematical Geology* 12(2): 85.

Goodchild, M.F., and D.M. Mark. 1987. The fractal nature of geographic phenomena. *Annals of the Association of American Geographers* 77(2): 265–278.

Hjelmfelt Jr., A.T. 1988. Fractals and the River-Length Catchment-Area Ratio. *Water Resources Bulletin* 24(2): 455–459.

Huang, J., and D.L. Turcotte. 1989. Fractal mapping of digitized images: Application to the topography of Arizona and comparisons with synthetic images. *Journal of Geophysical Research* 94(B6): 7491–7495.

Hutchinson, M.F. 1989. A New Procedure for Gridding Elevation and Stream Line Data with Automated Removal of Spurious Pits. *Journal of Hydrology* 106: 211–232.

Jenson, S.K. 1985. *Automated derivation of hydrologic basin characteristics from digital elevation model data.* Proceedings of Auto-Carto 7—Digital Representation of Spatial Knowledge, 301–310. Washington D.C.

Jenson, S.K. 1991. Applications of Hydrologic Information Automatically Extracted from Digital Elevation Models. In: *Analysis and distributed modeling in hydrology,* eds. Beven, K.J. and Moore, I.D. Terrain, 35–48. Chichester, U.K.: John Wiley and Sons.

Jenson, S.K., and J.O. Dominique. 1988. extracting topographic structure from digital elevation data for geographic information system analysis. *Photogrammetric Engineering & Remote Sensing* 54(11): 1593–1600.

La Barbera, P.L., and R. Rosso. 1989. On the fractal dimension of stream networks. *Water Resources Research* 25(4): 735–741.

Lee, J., K. Snyder, and P.F. Fisher. 1992. Modeling the effect of data errors on feature extraction from digital elevation models. *Photogrammetric Engineering & Remote Sensing* 58: 1461–1467.

Lee, J., and C.-J. Chu. 1996. Spatial structures of digital terrain models and hydrological feature extraction. *IAHS Publ.* No. 235.

Martz, L.W., and J. Garbrecht. 1992. Numerical definition of drainage network and subcatchment areas from digital elevation models. *Computers & Geosciences* 18(6): 747–761.

Moore, I.D., and R.B. Grayson. 1991. Terrain-based catchment partitioning and runoff prediction using vector elevation data. *Water Resources Research* 27(6): 1177–1191.

O'Callaghan, J.F., and D.M. Mark. 1984. The extraction of drainage networks from digital elevation data. *Computer Vision, Graphics and Image Processing* 28: 323–344.

Qian, J., W. Ehrich, and J.B. Campbell. 1990. DNESYS—An expert system for automatic extraction of drainage networks from digital elevation data. *IEEE Transactions on Geoscience and Remote Sensing* 28: 29–44.

Quinn, P., K. Beven, P. Chevallier, and O. Planchon. 1991. The prediction of hillslope flow paths for distributed hydrological modeling using digital terrain models. In *Terrain analysis and distributed modeling in hydrology*, ed. K.J. Beven, and I.D. Moore, 63–83. Chichester, U.K: John Wiley and Sons.

Robert, A., and A.G. Roy. 1990. On the fractal interpretation of the mainstream length-drainage area relationship. *Water Resources Research* 26(5): 839–842.

Tarboton, D.G., R.L. Bras, and I.R. Iturbé. 1988. The fractal nature of river networks. *Water Resources Research* 24(8): 1317–1322.

Tarboton, D.G., R.L. Bras, and I.R. Iturbe. 1989. Scaling and elevation in river networks. *Water Resources Research* 25(9): 2037–2051.

Tarboton, D.G., R.L. Bras, and I.R. Iturbe. 1991. On the extraction of channel networks from digital elevation data. In *Terrain analysis and distributed modeling in hydrology*, ed. K.J. Beven, and I.D. Moore, 85–104. York: John Wiley, New.

Tarboton, D.G., and M.E. Baker. 2008. Towards an algebra for terrain-based flow analysis. In *Representing, modeling and visualizing the natural environment: Innovations in GIS 13*, ed. N.J. Mount, G.L. Harvey, P. Aplin, and G. Priestnall, 167–194. Boca Raton, FL: CRC Press.

Tesfa, T.K.David, G. Tarboton, Daniel W. Watson, Kimberly A.T. Schreuders, Matthew E. Baker, and Robert M. Wallace. 2011. Extraction of hydrological proximity measures from DEMs using parallel processing. *Environmental Modelling and Software* 26(12): 1696–1709. ISSN 1364-8152.

Vieux, B.E. 1988. *Finite Element Analysis of Hydrologic Response Areas Using Geographic Information Systems*. Department of Agricultural Engineering, Michigan State University. A dissertation submitted in partial fulfillment for the degree of Doctor of Philosophy.

Vieux, B.E. 1993. DEM aggregation and smoothing effects on surface runoff modeling. *ASCE, Journal of Computing in Civil Engineering, Special Issue on Geographic Information Analysis* 7(3): 310–338.

Vieux, B.E., and N.S. Farajalla. 1994. Capturing the essential spatial variability in distributed hydrological modeling: hydraulic roughness. *Journal of Hydrological Processes* 8: 221–236.

Vieux, B.E., and S. Needham. 1993. Nonpoint-Pollution Model Sensitivity to Grid-Cell Size. *Journal of Water Resources Planning and Management* 119(2): 141–157.

Vieux, B.E., C. Chen, J.E. Vieux, and K.W. Howard. 2003. Operational deployment of a physics-based distributed rainfall-runoff model for flood forecasting in Taiwan. In: Proceedings, Weather *Radar Information and Distributed Hydrological Modelling*, IAHS General Assembly at Sapporo, Japan, July 3–11, eds. Tachikawa, B. Vieux, K.P. Georgakakos, and E. Nakakita, IAHS Red Book Publication No. 282: 251–257.

Wilson, J.P. 2012. Digital terrain modeling. *Geomorphology* 137(1): 107–121.

Chapter 8
Distributed Precipitation Estimation

Distributed Model Input

Abstract Historically, rainfall data for hydrologic applications have been obtained from a sparse network of rain gauges. A gauge samples rain at distinct points and therefore may not accurately reflect the spatial distribution of rainfall, especially from convective storms. Interest in using radar estimates of rainfall in distributed modeling comes from the desire to reduce errors due to imprecise knowledge of rainfall distribution in time and space. Traditionally, point estimates of rain gauge accumulations are distributed in space over the river basin by some means, such as Thiessen polygon, inverse distance weighting, and kriging. This chapter treats the topic of using radar and rain gauge networks to measure spatially distributed precipitation.

8.1 Introduction

Figure 8.1 shows cumulonimbus clouds rising over Oklahoma at approximately 9:00 pm, 16 June 2000 and the corresponding radar reflectivity at 8:57 pm is shown in Fig. 8.2. Reflectivity depends on the distribution of raindrop sizes with higher reflectivity (>35 dBZ—shown in black) representing more intense rainfall. Radar is an important source of spatially and temporally distributed rainfall data for hydrologic modeling. This chapter deals with the use of weather radar in hydrology, with primary emphasis at the river basin scale. Though this material relies heavily on the NEXRAD radar network deployed by the US National Weather Service, basic concepts may be applicable to other radars.

8.2 Rain Gauge Estimation of Rainfall

A network composed of many gauges is preferred over a single rain gauge for characterizing precipitation over a watershed. Such a network can be used alone or in conjunction with radar to measure spatially variable rainfall. The network

© Springer Science+Business Media Dordrecht 2016
B.E. Vieux, *Distributed Hydrologic Modeling Using GIS*,
Water Science and Technology Library 74, DOI 10.1007/978-94-024-0930-7_8

Fig. 8.1 Cumulonimbus clouds during a summer storm near Oklahoma city, 16 June 2000 at 9:00 pm

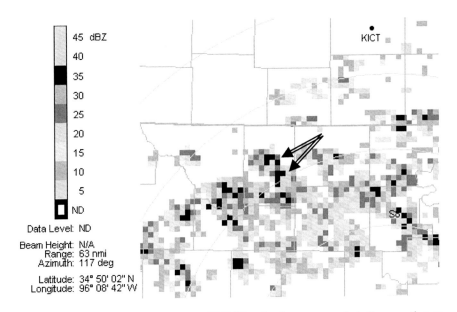

Fig. 8.2 Composite reflectivity from the NEXRAD radar that corresponds to the same time as when the photograph was taken

characteristics of density, recording method, telemetry, precision and type of equipment affect the accuracy, and representativeness of the data collected by the network. Storm type, convective or stratiform, and climatic factors also enter into planning a network. Intense precipitation often occurs over small areas, resulting in the most intense parts of the storm not being recorded by a gauge network. A convective storm event may pass through a gauge network and only affect a small number of gauges because of the limited spatial extent of the event.

The effect on rainfall measurement error caused by gauge network density relative to drainage area has been empirically determined from an early study on rainfall variability in the Muskingum River basin located in Ohio (U.S. Department of Commerce 1947). Approximately, four times the average density of gauges is required to reduce the standard error of measurement from 15 to 10 % based on the data from the Muskingum River basin in Ohio (see, U.S. Army Corps of Engineers 1994). Empirical gauge network density recommendations made by the U.S. Army Corps of Engineers (1996) relate the number of gauges, N_g, to the area modeled as:

$$N_g = A^{0.33} \tag{8.1}$$

where A is the watershed area in mi^2 ($2.59 \ km^2 = 1 \ mi^2$). The number of gauges for a range of watershed areas computed using Eq. 8.1 is shown in Table 8.1. The number of gauges recommended by this equation is quite small when compared to typical urban rain gauge catchments where densities may be as great as one gauge per 10–20 km^2. The minimum number of gauges implied by Eq. 8.1 is one for a watershed of 2.6 km^2 ($1 \ mi^2$). For a catchment of 100 km^2, Eq. 8.1 would suggest 3.3 gauges with a density of one gauge per 29.9 km^2.

Basing the number of gauges solely on a watershed area does not take into account the spatial variability of the rainfall and the timescale of rainfall (hourly, daily, or monthly) that must be captured by the gauge network.

The accuracy of a rain gauge network is deemed sufficient if it accurately measures rainfall over an area at required timescales, e.g., 5-min, hourly, or daily. From the standpoint of measurement theory for any random variable, the recommended number of samples is around 10–15. The origin of this recommendation stems from the diminishing reduction in standard error as the number of gauges is increased. The standard error of the mean, σ_{err}, may be used to estimate the closeness of the sample mean to the true mean for independent measurements:

$$\sigma_{err} = \frac{\sigma_s}{\sqrt{n}} \tag{8.2}$$

Table 8.1 Gauge number and density based on the drainage area

Number	Area (mi^2)	Area (km^2)	Density (km^2/gauge)	Density (mi^2/gauge)
1.0	1.0	2.6	2.6	1.0
2.0	8.2	20.9	10.5	4.1
4.0	66.7	170.9	42.7	16.7
8.0	545.2	1395.8	174.5	68.2

where σ_s is the standard deviation; and n is the number of independent observations. Figure 8.3 is a theoretical plot of standard error as a function of the number of gauges and the standard deviation of the rainfall storm total ranging from 5 to 50 mm. From the family of curves, beyond ten gauges, the standard error does not decrease significantly as more gauge observations are added. While Eq. 8.2 does not account for autocorrelation between gauges nor the size of the area covered, it does demonstrate the diminishing improvement in accuracy achieved by adding gauges to an area. Duration of the rainfall is indirectly taken into account in Eq. 8.2 because variance scales with depth and therefore duration of the rainfall accumulation.

If the precipitation is expected to have a standard deviation of 50 mm and we wish to approximate the population mean to within 10 mm, then the number of gauges required by Eq. 8.1 is 25. Correspondingly, for standard deviations of 10, 20, 30, and 40 mm, the number of gauges would have to be 2, 5, 10, 17, and 26 to achieve a standard error of the mean that is less than 10 mm. To capture the rainfall variability at the scale of a single event, more gauges are likely needed than is suggested by Fig. 8.3 or Eq. 8.2.

As noted, one difficulty with Eqs. 8.1 or 8.2 is that the spatial correlation of the rainfall field is unaccounted for, nor is the associated temporal aggregation, i.e., at hourly, daily, monthly, or annual periods. The standard error of the mean may serve as an estimate but does not consider the need for redundancy caused by gauge malfunction, limited spatial extent of the storm, and other factors that reduce reliability of a gauge network in operation. Because rainfall is spatially correlated, measurements obtained from a given set of gauges are likely not statistically independent.

The density of the rain gauge network required depends on the time scale of interest, i.e., event-, monthly, or annual-accumulations. The time scale of interest is linked to the spatial scale due to the autocorrelated nature of precipitation. Lebel and Le Barbé (1997) investigated how accurate seasonal rainfall accumulations are over a region in Niger. Using a geostatistical framework, this analysis identified the spatial autocorrelation of seasonal and event-scale accumulations. The region considered was the 1-square degree of the HAPEX-Sahel experiment (Goutorbe

Fig. 8.3 Standard error of rainfall measured by a number of rain gauges in a network

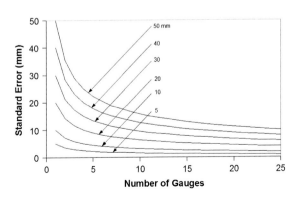

et al. 1994). This field experiment was designed to gather and analyze land-surface-atmosphere water and energy fluxes in a semi-arid environment in West Africa.

One component of the HAPEX-Sahel experiment was to characterize hydrologic fluxes described in Lebel and Le Barbé (1997) and Lebel et al. (1997 and 2009) in order to evaluate the accuracy of rain gauges used to validate satellite and radar estimates of rainfall. The goal of the HAPEX and other campaigns in this region was to characterize long-term climatic impacts on water resources. The rain gauge network density of 10 gauges per 10,000 km^2 reduced the estimation uncertainty for monthly rainfall to less than 10 % and less than 3 % for seasonal rainfall. Depending on the application, such precision may be considered sufficient for satellite rainfall algorithms at the GCM scale. For areas down to 1,000 km^2, the number of gauges is more important than density when considering monthly and seasonal estimates. For areas smaller than 1,000 km^2, the density should be such that the spacing is roughly half the decorrelation length, which is the distance beyond which measured values are no longer spatially correlated. For the region studied in Niger, the spacing is recommended to be less than 15 km (half of 30 km). If we enlarge our time domain of interest from hourly to daily, 10-day intervals, or monthly accumulations, the network can be less dense and still capture the spatial distribution of precipitation at longer timescales. Therefore, the density of a rain gauge network required to resolve the spatial variability of precipitation is relative to the event time scale, spatial autocorrelation of the aggregated rainfall and the area to be covered.

The spatial variability of rainfall plays an important role in the process of surface runoff generation. Faurès et al. (1995) examined how various rainfall measurement uncertainties and spatial rainfall variability affect runoff modeling for a small catchment. They found that runoff model runs simulated with data from a variable number of recording gauges demonstrated that the uncertainty in runoff estimation is strongly related to the number of gauges. Modeling of small events suffered greater relative variations than modeling of larger storms due to the larger relative percentage of measurement error. In a region characterized by convective thunderstorms, Goodrich (1990) noted significant differences in rainfall at two rain gauges separated by 300 m. Gradients in the same watershed, Walnut Gulch Arizona, ranged from 4 to 14 % in a 100 m distance. When these rainfall measurements were used as inputs to a distributed rainfall-runoff model on three small catchments, significant sensitivity of the model to spatial variability of rainfall input was noted (Goodrich 1990; Goodrich et al. 1995; Faurès et al. 1995 and Morin et al. 2006).

In regions where thunderstorms prevail, dense rain gauge networks are necessary to capture the significant spatial and temporal variability typical of convective storm events. Using one of the densest gauge networks found in the literature (14 rain gauges in a 36 ha catchment), Ambroise and Aduizian-Gerard (1989) noted that significant altitudinal rainfall variability was affected by topographically controlled wind direction in the mountain catchment. Second, they noted that systematic errors in rain gauge measurement can be due to water loss during the measurement, adhesion loss on the surface of the gauge and raindrop splash from the collector.

Third, they found that random errors, resulting from imprecision in measurement, are usually small and compensating. A more important error was found to be related to the nature of the sample location where rainfall catch variability was influenced by local topography. These rainfall measurement errors caused by topographic effects contain both systematic and random errors influenced by wind speed and direction. Emmanuel et al. (2012, 2015) examined the small-scale variability of rainfall from a geostatistical approach. They found four different types of rainfall events with distinctly different semivariograms. The range parameter of the semivariogram governing de correlation in time and space was longer for homogeneous rain events compared to shorter range parameters for small intense storms.

Spatial variability of rainfall and associated measurement accuracy of a rain gauge network was investigated by Vieux and Pathak (2007) and Pathak and Vieux (2008) for south Florida. Any rain gauge network is expected to accurately sample a temporally varying rainfall field that is spatially correlated. The measurement accuracy of the rain gauge network and its representativeness of rainfall over the target area are guiding principles for the design and optimization of the network. Because the network is used for more purposes than solely for climatology, e.g., operation of water control structures, the resulting configuration is different from having only a few gauges that are used to represent climatologically homogeneous areas. The existing gauge network consisted in 287 active gauges at the time of the study in 2006. The goal of the redesigned gauge network is to serve several purposes that are more demanding in terms of gauge density and distribution. Identification of an optimal number of gauges of variable density is based on accuracy requirements and the necessity to capture the spatial and temporal distribution of rainfall at hourly and daily intervals.

Estimation of the covariance structure (17,930 mi^2) area was accomplished using radar and rain gauge observations. The approach used analysis blocks (20 × 20 km) sufficiently large to encompass the range of spatial correlation but sufficiently small to capture climatological gradients near coastal areas and inland water features. The 15-min time interval measured by radar was first detrended within each analysis block. Then its covariance structure was determined as a correlogram and semivariogram to gain an understanding of the length scale beyond which the rainfall becomes decorrelated. The spatial correlation coefficient as a function of separation, ρ_h, is computed as:

$$\rho_h = e^{-\theta h} \tag{8.3}$$

where θ is a decay coefficient that models the amount of correlation with separation distance, h. Similarly, the exponential semivariogram and R the sill parameter can be calculated as:

$$\gamma_h = \frac{1}{2}\sigma_o^2\left(1 - e^{\left(\frac{-3h}{R}\right)}\right) \tag{8.4}$$

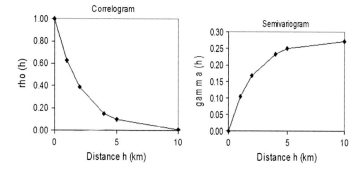

Fig. 8.4 Representative correlogram (*left*) and semivariogram (*right*) showing decorrelation after 5 km distance

where R is the range with units of distance; h is the lag distance and, σ_o^2, is the sill, which is also the point variance of the rain gauge. Figure 8.4 shows a representative example for an analysis block. On the left is the correlogram (Eq. 8.3) and on the right the semivariogram for a range parameter, $R = 5$ km, and with zero sill for comparative purposes.

The correlation coefficient depends on the range parameter established by radar and the spacing, h, which corresponds to the desired average spacing of N gauges per analysis block. From the radar-derived range parameter, R and with average inter-gauge spacing, h, the correlation between any given number of gauges within each block can be established. The spatial correlation of gauges reduces the effective number of gauges because they are not independent. Haan (2002, p. 290) defined the average interstation correlation and resulting effective number of gauges. The effective number of gauges can be computed from the correlation coefficient for any given inter-gauge distance. The effective number of gauges, N_e, is computed as:

$$N_e = \frac{N}{1 + \rho_h(N - 1)} \tag{8.5}$$

where N is the given number of gauges and ρ_h is the correlation coefficient. The standard error of the mean computed for the effective gauge number is:

$$SE = \sqrt{\frac{\sigma_o^2}{N_e}} \tag{8.6}$$

The median range parameter computed from 1998–2005 is shown in each analysis block covering the lower half of the state from offshore to interior portions of the district. Whereas, similar range parameters were found for hourly and daily intervals, the sill or point process variance differed greatly. The median of the daily range parameters was 6.34 km and the hourly is 6.49 km. From mapped gauge variance (not shown), the median point variance of the daily interval was $\sigma_o^2 = 177.4$ mm^2

6.1	6.1	6.9	6.6	6.8	6.7	6.1	6.6	6.7	6.0	5.4	6.1	5.7	5.4	5.8	6.2	6.0	6.2
5.8	6.3	6.8	6.7	6.2	6.7	6.4	7.1	6.4	6.2	6.2	6.4	5.4	5.7	5.8	5.1	5.7	5.9
5.8	6.5	6.3	6.5	6.6	6.6	6.9	5.9	6.6	6.1	6.1	5.3	5.7	5.7	6.0	5.6	6.1	5.9
6.2	6.0	6.4	5.9	6.9	6.4	5.7	6.6	5.7	6.2	5.6	5.2	5.8	4.9	5.7	5.6	5.5	6.6
6.3	6.5	6.7	6.1	6.1	6.1	6.4	6.3	5.8	5.9	5.6	5.8	6.4	5.2	5.0	5.2	6.0	6.9
6.7	6.5	6.4	6.7	6.1	6.7	6.4	6.5	6.0	6.7	6.6	6.4	6.2	5.8	5.5	6.0	5.9	7.1
6.9	6.4	6.5	6.2	6.3	6.1	6.3	6.4	6.7	6.6	6.0	6.4	6.1	5.3	5.7	5.9	6.1	6.5
5.7	6.4	6.3	6.4	6.3	6.4	6.0	6.7	6.4	6.9	6.4	6.4	6.6	5.4	5.4	5.2	5.8	6.2
5.7	6.3	6.5	6.5	5.8	6.4	6.7	6.6	6.8	6.7	6.6	6.2	6.5	6.1	5.6	6.1	5.6	5.7
5.1	5.8	6.2	6.5	6.2	6.1	6.6	6.6	6.7	6.3	6.7	6.5	6.8	6.7	6.1	6.2	6.0	5.9
5.9	6.3	6.7	5.8	6.5	7.0	6.5	6.6	6.2	6.8	6.9	7.5	6.9	6.6	6.0	5.7	5.6	5.5
5.5	5.3	5.6	6.8	6.4	6.2	6.7	6.5	6.3	6.7	6.7	6.7	6.5	6.5	6.1	5.9	6.0	6.0
5.6	6.1	6.4	6.3	5.9	6.8	6.6	6.2	6.6	6.4	7.1	6.8	7.2	6.4	6.3	5.9	5.6	5.2
5.7	6.0	5.9	6.3	6.4	6.7	6.5	7.4	6.5	5.8	6.6	6.3	6.7	6.6	6.0	6.4	6.2	5.5
6.0	5.4	5.7	6.8	6.2	6.8	6.7	6.6	6.4	6.9	6.5	6.5	6.9	6.8	5.7	6.6	5.8	6.1
6.0	5.7	6.5	6.4	5.7	5.8	6.2	6.7	6.4	6.2	6.8	6.8	7.3	7.1	6.3	5.9	6.1	6.2
5.9	5.6	6.3	6.4	5.9	7.3	6.6	6.7	6.9	6.4	6.4	6.8	6.1	6.3	5.9	5.7	6.2	6.0
6.1	5.8	6.5	6.0	6.5	7.0	7.1	6.4	6.2	6.4	6.8	6.4	6.6	6.5	9.2	5.7	6.2	6.2
4.9	5.4	5.5	6.5	5.9	6.8	6.7	6.5	6.4	6.9	7.2	6.9	7.6	7.5	5.9	5.6	6.2	5.8
5.1	5.8	5.9	6.6	5.8	6.6	6.3	6.7	6.5	6.8	7.4	6.1	6.4	6.4	6.0	6.0	5.6	6.1
5.1	5.5	5.5	6.4	6.8	6.8	6.9	6.3	7.1	7.2	6.7	6.4	6.2	6.5	6.1	6.3	6.1	6.1
5.8	6.0	5.4	5.9	6.4	6.2	6.4	6.0	6.0	6.8	7.0	6.1	6.2	6.1	6.0	5.6	5.9	6.0

Fig. 8.5 Median wet season range parameter, R in km, for each 20 × 20-km analysis block (1998–2005)

(0.275 in.^{2}) and hourly variance, $\sigma_o^2 = 24.84$ mm^2 (0.0385 in.^{2}) is assigned to each analysis block. Figure 8.5 shows the median wet-season range parameter for a daily time interval in each analysis block in south Florida (dark gray).

By way of an example, the standard error of the mean rainfall $N_e = 3.90$ gauges. With four gauges instead of one, the standard error of the mean will be reduced by nearly 50 % in error, i.e., from 13.3 to 6.75 mm = 3.90 gauges.

Parameter	Daily	Hourly
R (km)	6.34	6.49
σ_o^2 (mm^2)	177.4	24.84

Daily interval—

$N = 4$ gauges
$h = (400/4)^{1/2} = 10$ km
$R = 6.34$ km
$\sigma_o^2 = 177.4$ mm^2
$\rho_h = e^{(-3*10/6.34)} = 0.00881$
$N_e = 3.90$
SE $= (177.4/3.90)^{1/2} = 6.75$ mm

For $N = 1$

SE $= (177.4/1)^{1/2} = 13.3$ mm

The number of gauges required to achieve a given standard error can now be computed for each of the analysis blocks seen in Fig. 8.6. The numbers of gauges shown achieve a SE = 5.08 mm (0.2 in.). The density is variable with a concentration of gauge numbers found closer to the coast where rainfall gradients produce the most variable rainfall primarily affected by high point variance.

Fig. 8.6 Optimal number of gauges for each 20×20 km analysis block that achieves a given standard error

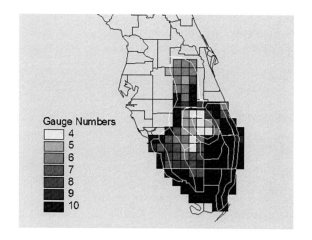

Table 8.2 Average network density for specified standard error

Characteristic	Standard error (mm)			
	5.08	6.35	7.62	8.89
Gauge numbers	1037	576	332	226
Area per Gauge	48	87	150.6	221
Spacing	7	9	12.3	15

Required gauge numbers can then be computed for a range of standard errors, together with the area per gauge. Table 8.2 shows the requisite network that achieves a given standard error in terms of numbers, average density, and spacing.

8.3 Radar Estimation of Precipitation

The NEXRAD (called WSR-88D) network of weather radars serves a wide range of weather-related applications for the US. The radar hardware characteristics and algorithm development determine its ability to provide useful hydrologic model input. Radar does not measure precipitation directly and relies on reflected signal strength from the distribution of raindrops in a given volume of the atmosphere. Reflectivity values are obtained by measuring the power of backscattered radiation. The radar reflectivity factor, Z (mm^6 m^{-3}) is given by:

$$Z = V^{-1} \sum_{i=1}^{n} N_i D_i^6 \qquad (8.7)$$

where N_i is the number of drops of a particular size; n is the number of raindrop size classes and D_i is the diameter (mm) of the ith drop size class. The drop size distribution can change dramatically within the storm, depending on the origin or genesis of the precipitation event. Rainfall rate and reflectivity are related because both depend on the drop size distribution represented in Eq. 8.7 by the number of drops of a given size.

The radar equation relates power, P, to characteristics of the radar and to characteristics of the precipitation targets and is given by:

$$P = \frac{C * L * Z}{r^2} \qquad (8.8)$$

where P denotes radar-measured power (watts); C is a constant that depends on radar design parameters such as power transmission, beam width, wavelength, and antenna size; L represents attenuation losses; Z is the radar reflectivity factor (mm^6 m^{-3}); and r is the range (km).

Because the radar is measuring a surrogate, reflectivity, the rainfall estimate is likely to be in error to a greater or lesser degree depending on the Z-R relationship being used. The reflectivity, Z, is related to rainfall rate, R (mm/h), by the following:

$$Z = \alpha R^{\beta} \qquad\qquad (8.9)$$

where Z is reflected power in units of $mm^6 \ m^{-3}$; R is rainfall rate in mm/h; and α and β are coefficients derived empirically to represent a drop size distribution. For convective storms, the Z-R relationship of $Z = 300R^{1.4}$ is recommended, while rainfall events that are driven by warm processes, typical of tropical air masses, are better represented by $Z = 250R^{1.2}$ (Rosenfeld et al. 1993). Doviak and Zrnic (1993) described the basic process of converting reflectivity to rainfall rate using Z-R relationships and the factors that contribute to uncertainty in precipitation estimation.

As mentioned above, the Z-R relationship used to convert reflectivity to rainfall rate is an empirical relationship that depends on the drop size distribution. This distribution changes throughout the storm and depends on the origin and evolution during the storm event of the precipitation-producing mechanism. To overcome the systematic errors inherent in the Z-R relationship, calibration with rain gauges can be performed in real-time or in post-analysis mode. This procedure described below consists in comparing accumulations between radar and gauge.

Morin et al. (1995) observed that good temporal (5 min) and spatial (1-km^2) resolution of rainfall is attainable over large areas using a meteorological radar. The large variability between the precipitation echo intensity ($mm^6 \ m^{-3}$) and the rain intensity ($mm \ h^{-1}$) can cause significant errors in rainfall estimation. This may be overcome by adjusting the Z-R relationship using rain gauge measurements. Rosenfeld et al. (1994, 1995a, b) developed and applied the window probability matching method (WPMM). The WPMM method better accounts for much of the variation in the Z-R relationship significantly improving the accuracy of the radar estimated rainfall. The WPMM approach relies on matching probabilities of radar observed reflectivity (Z) to rain-gauge measured rain intensity, R, taken from small 'windows' centered over the gauges, which have been objectively classified into different rain types. Applying this method to radar measurements over several catchment areas in Central Israel and comparing daily rain-gauge measurements with radar rainfall estimates demonstrated good agreement (Morin et al. 1995). The WPMM accounts for variation in drop size distributions unlike other methods that optimize temporally and/or spatially averaged values.

Even if a Z-R relation perfectly represents the relationship between rainfall rate and reflectivity, calculated differences between radar and ground measurements may still result from:

- Synchronization and directional errors of the radar with respect to ground location
- Large variations/gradients in rain intensity within the measured atmospheric volume
- Sampling error resulting from point measurement of the rain gauge
- Using a suboptimal time interval for rain-gauge intensity integration.

The most common adjustment of radar to gauge accumulations is performed using the method of Wilson and Brandes (1979). This adjustment involves a multiplicative factor that removes the bias between the radar and gauge accumulations. Before we address bias correction, the reason for its occurrence, i.e., differences between the *assumed* and *actual* rainfall drop size distribution, is discussed as follows.

8.3.1 Rainfall Drop Size Distributions

Understanding how radar can be used to measure rainfall requires a probabilistic understanding of rainfall. Instead of measuring some depth, radar measures a surrogate measure, reflectivity. We define rainfall as a range of drop sizes that follow some probability distribution function. The most common probability density function (PDF) is the exponential distribution with two parameters, a mean or median drop size and number of drops per unit volume as the drop size approaches zero. The number of drops for all sizes per unit volume, N, is defined as:

$$N = N_o e^{-\Lambda D} \qquad (8.10)$$

where N_o is the number of drops of size zero (as the diameter approaches zero, it converges to N_o); Λ is the mean drop size with units of $1/D$ (mm^{-1}). The mean drop size, Λ, may be written in terms of the median drop size, D_o, with $\Lambda = 3.67/D_o$. Writing Eq. 8.10 in integral form results in a relationship for reflectivity in terms of the exponential PDF expressed by Eq. 8.10. Substituting Eq. 8.10 into Eq. 8.9, the reflectivity factor, Z, becomes:

$$Z = \int_0^\infty N_o e^{-\Lambda D} D^6 dD \qquad (8.11)$$

where all terms are as defined before. In the practical application of Eq. 8.11, the maximum drop size is rarely more than 3.5 or 4 mm. Therefore, the integral in Eq. 8.11 should be integrated from zero to the maximum drop size rather than infinity. For this reason, the *incomplete* gamma function should be used. Evaluation of the incomplete gamma function requires numerical integration or a mathematical table of values. Recognizing that the *complete* gamma function is a simple factorial, where $\Gamma(7) = 6!$, we can easily integrate Eq. 8.11 with limits between zero and infinity. The reflectivity factor, Z, in this case becomes:

$$Z = N_o \frac{\Gamma(7)}{\Lambda^7} = N_o \frac{6!}{(3.67/D_o)^7} \qquad (8.12)$$

The rainfall rate calculation is approached in a manner similar to the reflectivity factor. If we know the distribution and fall velocity of each drop size, then we can compute the rainfall rate, R (mm h^{-1}). The mass of a water drop, $m(D)$, is:

$$m(D) = (\pi/6)D^3 \rho_w \tag{8.13}$$

where ρ_w is the density of water. The time arrival of the mass is determined by the fall velocity, $w(t)$, of each drop size, which can be estimated as:

$$w(t) = (386.6)D^{0.67} \tag{8.14}$$

The rainfall rate expressed as mass per unit time, R, is:

$$R = \int_0^\infty m(D)N(D)w(D)dD \tag{8.15}$$

Dividing R in Eq. 8.15 by the density of water, ρ_w, we obtain the rainfall rate expressed as a depth over a unit area. Substituting Eqs. 8.10, 8.13, and 8.14 into Eq. 8.15 yields the rainfall rate, R, as:

$$R = (\pi/6) \int_0^\infty D^3 N_o e^{-\Lambda D} 386.6\, D^{0.67} dD \tag{8.16}$$

Integration of Eq. 8.16 using the gamma function along with $\Lambda = 3.67/D_o$, results in:

$$R = N_o D_o^{4.667}/4026 \tag{8.17}$$

As an example, using the parameters of the exponential distribution that fit the Marshall-Palmer (M-P) distribution (Marshall and Palmer 1948):

$$N_o = 8000 \text{ drops}/(\text{mm}^{-1}\text{m}^3)$$
$$D_o = 2.42 \text{ mm}$$

Substituting these values of N_o and D_o into Eq. 8.17 yields the rainfall rate:

$$R = 114 \text{ mm/h}$$

Writing the reflectivity factor, Z, defined by Eq. 8.12 in a more convenient form, we obtain:

$$Z = 0.080 \, N_\text{o} D_\text{o}^7 \tag{8.18}$$

Substituting the same M-P DSD parameters ($N_\text{o} = 8000$, and $D_\text{o} = 2.42$ mm) into Eq. 8.13, we obtain the reflectivity factor, Z = 311091. The reflectivity factor is usually written in terms of a decibel of reflectance (dBZ), which by definition is:

$$10 * \log\,(311091) = 54.9 \text{ or } \sim 55\text{dBZ}$$

By adopting a probabilistic view of rainfall defined by the DSD, reflectivity and rainfall rate may be related. Though not as apparent, the reader can verify that different parameters of the DSD can yield the same Z- or R-values. Nonuniqueness is one reason for using rain gauges to adjust radar estimates of rainfall from the single parameter, reflectivity. Multiple parameters can be measured by polarizing the microwave beam and then deriving parameters based on horizontal and vertically polarized reflectivity, see for example, Doviak and Zrnic (1993) and references therein.

8.3.2 Z-R Relationships

As seen above, the DSD determines both the reflectivity factor *and* rainfall rate, forming the basis for a Z-R relationship. Z-R relationships may be formed by combining Eqs. 8.17 and 8.18 with appropriate parameters of the DSD, resulting in:

$$Z = 228R^{1.5} \tag{8.19}$$

This is approximately the same as the M-P relationship or $Z = 200R^{1.6}$ but differs because of the integration limits in Eqs. 8.11 and 8.16. In general, the values of N_o, D_o and maximum drop size depend on the type of rainfall process, storm evolution and other factors (Doviak and Zrnic 1993) and are not known *a priori*. Downdrafts and updrafts affect the fall velocity of drops depending on drop size and strength of the local winds produced during a storm. Rainfall rates estimated by means of a Z-R relationship generally do not account for local wind effects on fall velocity. Even after bias removal, which adjusts the multiplicative coefficient α in Eq. 8.9, random errors still exist. Figure 8.6 shows how the two parameters of the DSD combine to affect rainfall rate over a range of median drop sizes, D_o and number of drops, N_o. Drop sizes over 3 mm have only a small influence at high rainfall rates as shown in the graph by the 100 mm/h curve because of their low numbers.

When WSR-88D radars were first deployed in the mid-1990's, the system had only one 'standard' Z-R relationship that was used throughout the network. In some locales, the NWS has adopted the 'tropical' Z-R relationship, $Z = 250R^{1.2}$, which is more representative of warm tropical rainfall drop distributions (Rosenfeld et al. 1993). The tropical Z-R is representative of DSDs that tend to have a great number

Fig. 8.7 Rainfall rate determined by drop size and number of drops in the M-P DSD

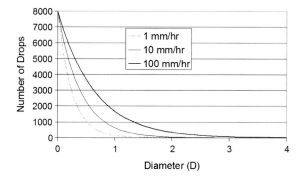

of small raindrops but produce copious amounts of rainfall. It is used operationally in radar installations impacted by tropical storms, e.g., along the Gulf Coast.

In effect, calibration is an adjustment of the multiplicative constant, and secondarily the exponent, in the Z-R relationship. During a major rainfall event in south-east Texas, October 1994, the WSR-88D radar at the Houston-Galveston (KHGX) underestimated the rainfall by as much as 50 % (NWS 1995). Vieux and Bedient (1998) found that the tropical Z-R ($Z = 250R^{1.2}$) relationship better characterized the October 1994 storm event. Using daily accumulations, the mean field bias (MFB) for this event ranged from 6 % underestimation on 17 October to 15 % overestimation on October 18. From this study, WSR-88D was found to be an accurate source of rainfall information provided that an appropriate Z-R relationship is used. Without bias correction, there was a 15 % overestimation on October 18 indicating that the Tropical Z-R was representative of the actual DSD.

A more convenient form of the Z-R relationship can be written with R as the dependent variable. The 'tropical' and 'standard' Z-R relationships may be written as:

Tropical Z–R relationship

$$R = \left[\frac{10^{\left(\frac{dBZ}{10}\right)}}{250}\right]^{(1/1.2)} = \frac{10^{\left(\frac{dBZ}{12}\right)}}{99.6} \qquad (8.20)$$

Standard Z-R relationship

$$R = \left[\frac{10^{\left(\frac{dBZ}{10}\right)}}{300}\right]^{(1/1.4)} = \frac{10^{\left(\frac{dBZ}{14}\right)}}{58.8} \qquad (8.21)$$

Any Z-R relationship may be similarly rearranged for computational convenience. To see the difference introduced by using one Z-R as opposed to another, we can substitute a range of dBZ values into Eqs. 8.20 and 8.21 and observe the difference in estimated rainfall rate, R (mm/h). The rainfall rates produced using these Z-R relationships are compared in Table 8.3 for a typical range of reflectivity (dBZ). Because of the logarithmic scale used to report reflectivity in dBZ, it is important to keep in

Table 8.3 Comparison of Z-R relationships for a range of reflectivity

Z	dBZ	Tropical R (mm/h)	Standard R (mm/h)
100	20	0.47	0.46
316	25	1.22	1.04
1000	30	3.17	2.36
3162	35	8.29	5.38
10000	40	21.63	12.24
31623	45	56.46	27.86
100000	50	147.36	63.40

mind the corresponding rainfall rates. An increase of just 5 dBZ (e.g., from 35 to 40 dBZ) results in a more than threefold increase in Z, which translates into 21.63 mm/h compared with 8.29 mm/h in the tropical Z-R relation. Small increases in detected reflectivity (dBZ) represent large increases in rainfall rate.

The large differences between a tropical and standard Z-R relationship are evident in the last two columns of Table 8.3 above. If a reflectivity of 45 dBZ was detected, the tropical relation would estimate the rainfall rate to be 56.46 mm/h, compared with 27.86 mm/h for the convective Z-R relationship. Operationally, it is difficult to choose the appropriate Z-R relationship. NEXRAD radar operators may switch from one Z-R to another to better match observations. However, since the underlying physics of the rainfall-producing process governs the drop size distribution, adopting an appropriate Z-R relationship is difficult. A high reflectivity cap is also applied to reduce the influence of water-coated hydrometeors like hail. This is an adaptable parameter at the radar installation, which is usually set to 103.8 mm/h, which corresponds to 53 dBZ using the standard convective Z-R relation.

Assuming any Z-R relationship is essentially assuming a given median drop size and number of droplets per unit volume. Occasionally, the Z-R relationship will be correct, i.e., the assumed Z-R relationship actually represents the DSD that occurred. More often than not, the assumed Z-R does not represent the actual DSD of any given storm and requires adjustment for a storm period or during the storm. Other radar characteristics described below can affect rainfall estimation as well.

8.3.3   Radar Power Differences

The strength of returned power reflected from raindrops determines the rainfall rate. Two radars should measure the same reflectivity at a given location in the atmosphere under ideal conditions. The power produced by each radar according to Eq. 8.4 should be the same if the same equipment is used, e.g., two WSR-88D radars. However, the power transmitted and received is rarely the same due to small differences in transmitter and receiver parts and due to differences along the path that the beam follows in the atmosphere. If a radar is underpowered in relative or absolute terms, the rainfall rate will be underestimated. The NWS attempts to calibrate the WSR-88D radar hardware to bring neighboring radars to within 1 dB

Table 8.4 Rain rate errors associated with differences in radar power

Power Difference (dB)	$Z = 300R^{1.4}$	$Z = 250R^{1.2}$
−4	52 % of actual Multiply by 2.0	44 % of actual Multiply by 2.25
−3	60 % of actual Multiply 1.7	55 % actual Multiply by 1.8
−2	72 % of actual Multiply by 1.4	67 % actual Multiply by 1.5
−1	85 % of actual Multiply by 1.2	80 % of actual Multiply by 1.25
0	No error	No error
1	118 % of actual Multiply by 0.85	125 % of actual Multiply by 0.80
2	140 % of actual Multiply by 0.7	150 % of actual Multiply by 0.65
3	166 % actual Multiply by 0.6	183 % actual Multiply by 0.55
4	192 % actual Multiply by 0.5	225 % actual Multiply by 0.45

of each other. To gain an appreciation of how much power differences affect rainfall rate measurement, consider Table 8.4 adapted from Chrisman and Chrisman (1999). The radar power difference can be considered as a relative difference between two radars, or an absolute difference from a specified standard. In either case, the correction necessary to compensate for the power difference is shown for two of the Z-R relationships in common use by WSR-88D facilities. If two radars have a power difference of −2 dB, then the rainfall rate will be 72 % of the other using the convective Z-R relationship and 67 % of the other for the tropical relationship. Correction of this underestimation would require a multiplicative factor of 1.4 and 1.5 for the two Z-R relationships tabulated, respectively. Alternatively, the two radars could be calibrated so that they are in closer agreement. However, this procedure requires the owner of the radars to make technical modifications or adjustments that are beyond the control of the end user. The US NWS has scheduled hardware calibration that will eventually bring the WSR-88D network into closer agreement.

Because of these power differences, mosaicking radars that have not been calibrated can result in anomalous rain rates that are particularly evident in areas of overlap. Pereira et al. (1998) found artifacts and errors result when mosaics are produced by averaging overlapping rainfall estimates. Their analysis showed that hourly estimates in overlapping areas were in error by as much as 40 % due to the averaging of two different radar estimates. These errors can result from power differences, or other causes associated with radars that take measurements at different beam elevations in the atmosphere.

8.3.4 *Radar Bias Adjustment*

As with any measurement, both random errors and systematic errors (bias) may be present. However only systematic errors may be removed or mitigated by application of a correction factor. This correction is often termed calibration or adjustment, or bias correction. Bias adjustment and comparisons between gauge and radar go back to the 1970s (Wilson and Brandes 1979; Zawadski, 1973 and 1975). Removal of systematic errors in radar estimates using the multiplicative bias correction factor, F, is achieved by defining the ratio of the mean gauge and radar accumulations as:

$$F = \frac{\frac{1}{n}\sum_{i=1}^{n} G_i}{\frac{1}{n}\sum_{i=1}^{n} R_i} \qquad (8.22)$$

where G_i and R_i are the 'ith' gauge-radar pairs of accumulations and n is the number of pairs. When F in Eq. 8.22 is computed for a storm event total, G_i and R_i represent hourly, storm total, or another integration period for accumulating rainfall during the event. Operationally, application of this correction factor requires rain gauge measurements to be transmitted online. The NWS implementation of an online bias correction scheme was described by Seo et al. (1999) and Seo and Breidenbach (2002). In post-analysis, the application of the mean-field-bias correction factor, F, removes the systematic error (bias) from the radar. As Wilson and Brandes (1979) observed, combining radar and rain gauge measurements results in better precipitation estimates than can be obtained from either system alone.

Flood damages resulted from Tropical Storm Allison in the Houston Texas area in 2001. A radar-based flood alert system relied on radar data to produce warnings of impending damages within the Brays Bayou watershed comprising an area of 260 km² (Bedient et al. 2003). The storm total during a portion of Tropical Storm Allison, on 5 June 2001, is shown in Fig. 8.8. The standard deviation measured by 11 rain gauges in close proximity to this basin was 54.35 mm. The standard error of the mean achieved for this event using 11 rain gauges was 16.39 mm.

Adjusting radar using rain gauge accumulations using Eq. 8.22 relies on taking pairs of observations from gauges, G_i and from radar at the same location, R_i. Table 8.5 shows the results of comparing radar to gauge accumulations in Houston, Texas. The network consists in telemetered gauges operated at a local authority (HCOEM).

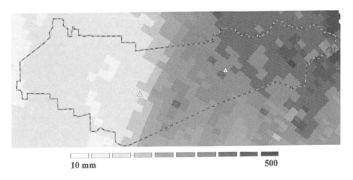

10 mm   500

Fig. 8.8 Storm total for Brays Bayou during Tropical Storm Allison, June 9, 2001

Table 8.5 Storm total gauge and radar for Tropical Storm Allison in Houston, TX

Gauge ID	Gauge (G_i) (mm)	Radar (R_i) (mm)	Adjusted Radar (R_i^*) (mm)
400	167.5	154.1	149.6
410	167.0	228.4	221.7
420	106.0	111.2	108.0
430	132.8	108.0	104.8
440	70.9	87.5	84.9
460	53.9	58.5	56.8
465/475	44.8	41.8	40.6
470	33.3	26.1	25.4
480	40.7	41.3	40.1
485	23.6	19.0	18.4
490	36.0	27.0	26.2
MEAN=	79.7	82.1	79.7
STDEV=	54.3	64.9	63.0
STERR=	16.4	19.6	19.0

The multiplicative bias correction factor, F, in Eq. 8.22 is computed as the ratio of the mean gauge and radar accumulations, $F = 79.7/82.1 = 0.97$, or nearly 1.0. Applying this correction factor the radar accumulations, R_i, results in the adjusted radar values in the last column. Note that once the bias correction factor is applied, the mean of the radar accumulations changes from 82.1 to 79.7 mm, which agrees with the mean of the gauge accumulations (79.7 mm). This example illustrates the mean field bias correction method. If this factor is allowed to vary in space, it is referred to as a local bias correction (Seo and Breidenbach 2002).

The average difference after bias removal is an indicator of the uncertainty of the radar rainfall estimate sampled over the rain gauge, which also is termed as *random error*. The average difference is defined as:

$$\overline{D} = \frac{100\,\%}{n} \sum_{i=1}^{n} \left| \frac{G_i - R_i^*}{G_i} \right| \tag{8.23}$$

The average difference after bias correction is 15.94 %, which indicates that the radar is within ±8 %. The average difference is only an indication of the random error measured by comparison to rain gauges. Gauge errors cause uncertainty, too. Errors in gauge accumulations can be caused by a variety of effects including electromechanical malfunction or wind effects. Tipping bucket gauges are known to under-report during heavy rainfall rates. Other uncertainty in the radar measurement likely exists that is not captured through comparison with point values measured by gauges. The benefit of combining radar and gauges is reduced systematic error in the radar-derived precipitation measurement. Removing the bias has a major influence on hydrologic predictions. Adjusting radar with gauges improves hydrologic predictions. Mimikou and Baltas (1996) compared the accuracy of radar with point rainfall values from gauges for flood forecasting. When bias-corrected radar data were input into a hydrologic model, the hydrograph rising limb and peak flow proved more accurate than the hydrographs produced from the rain gauge data alone. Another advantage of radar over rain gauge networks for rainfall estimation is the density of measurement. Vieux and Bedient (2004) demonstrated that hydrologic prediction accuracy improves when using bias-corrected radar in a distributed flood forecasting system. This type of evaluation was extended by Looper and Vieux (2012) and Looper et al. (2012) but focused on improvement achieved in flood forecasting accuracy by the use of gauge-adjusted radar rainfall (GARR) compared with rain gauge-only (RGO) rainfall as model input. Looper and Vieux (2012) found that the rain gauge density required to achieve accuracies equivalent to those obtained with GARR input was approximately five times denser and that the predictive streamflow accuracy using RGO input diminishes with rain gauge density about seven times faster than with the GARR input.

8.3.5 WSR-88D Radar Characteristics

The U.S. NWS along with other agencies deployed a network of weather surveillance radars for nationwide coverage. These radars are known as the WSR-88D and more commonly as NEXt generation RADar, or NEXRAD. Crum and Alberty (1993) described the WSR-88D system design and radar products. This radar is a Doppler radar with a 10-cm wavelength (S-band) transmitter that records reflectivity, radial velocity, and spectrum width. Compared with radars with shorter wavelength radars (X- or C-band), the S-band radar with a 10-cm wavelength suffers less attenuation in heavy rainfall making it useful for hydrologic applications. The WSR-88D is a volume-scanning radar, meaning that successive tilt angles are employed to cover large volumes of atmosphere out to 460 km for reflectivity and 230 km for precipitation, velocity, and spectrum width. Each volume scan starts at a tilt angle of 0.5°. The tilt angle of the radar beam is a principal reason why mosaicking two radars together will combine measurements taken at different elevations in the atmosphere.

Depending on the volume coverage pattern (VCP) used to identify different types of meteorological phenomena, the time taken to complete the volume scan varies, affecting the time step of rainfall inputs derived from the radar. Scanning strategies and development of new VCPs are continuing to change as the NWS makes operational changes to the system. Three weather conditions determine the number of sweeps/volume scan (one complete revolution at a particular tilt): 14 tilts during severe precipitation events, 9 tilts during non-severe precipitation events and 5 during clear air conditions. Correspondingly, the rates of data acquisition are one volume scan per 5, 6, and 10 min for VCP 11, VCP21, and VCP31/32, respectively. If the radar is operating with VCP 11, the temporal update of reflectivity is every 5 min, which means that the smallest time increment for input to a hydrologic model is 5 min. Radar scanning characteristics of the recorded reflectivity affect the intervals at which rainfall rates are updated. When used in hydrologic applications, the time intervals between recorded reflectivity have important consequences on model results, depending on the scale and application of the radar rainfall estimates.

New VCPs were added by NWS in 2004, which the radar operator may use to scan the atmosphere more quickly. This results in higher volume scanning rates and more tilt levels closer to the ground. VCP 12 will require 4.1 min to complete. While VCP 12 will have the same number of elevation scans as VCP 11, denser vertical sampling at lower elevation angles will provide a better vertical definition of storms in the lower atmosphere. Increased detection capabilities of radars impacted by terrain blockage will improve rainfall and snowfall estimates and result in more storms being identified and provide quicker updates. A consequence of the VCP used in operations is that derived rainfall rate estimates will have variable temporal update frequencies ranging from approximately 5 min.

Because individual radars in the WSR-88D radar network are not synchronized, a mosaic is generated once the radars have finished their individual volume scans. For this reason, mosaicked products are generally not available at short time intervals. During a storm event, the radar may switch from one VCP to another making the time intervals of the recorded data uneven. Post processing is required to sample this data into even increments, especially when a mosaic of more than one radar is required.

WSR-88D radars are often installed and operated remotely from NWS forecast offices. Figure 8.9 shows a WSR-88D radar installed near Norman, Oklahoma. The parabolic dish reflector of the radar is mounted inside the radome (white covering that envelops the antenna) to shield it from wind currents. This radar is one of approximately 160 WSR-88D radars deployed nationwide and overseas and is used by the NWS Radar Operations Center in Norman, Oklahoma for testing radar processing algorithms.

Operational and meteorological factors, such as transmitter power, signal attenuation, and raindrop size distribution affect the rainfall rates estimated by radar. Assessment of radar facilities and intercomparison for urban hydrologic applications was reported by Vieux and Vieux (2003), Vallabhenini et al. (2003), and Einfalt et al. (2004). There are several precipitation products with varying precision, spatial, and temporal resolutions. These products, as well as components comprising the precipitation processing system (PPS), are described in Fulton et al. (1998). The hydrologist is faced with two choices: (1) using products generated by the NWS, or (2) developing rainfall rates from reflectivity. The former choice is easier because processing has already been performed by the PPS. However, in this case, resolution and update intervals are fixed by the PPS. The latter choice is more difficult but avoids processing decisions made by the PPS, allowing more flexibility.

The NEXRAD system has evolved with recent enhancements that include dual polarization and other system improvements (Berkowitz et al. 2013). The U.S. Department of Commerce (2013, 2005, 2006a, b) provides a thorough description of the radar system, theory, operation and meteorological data products and processing.

Fig. 8.9 WSR-88 radar installation, Norman, Oklahoma

8.3.6 *WSR-88D Precipitation Processing Stream*

A brief description of the WSR-88D data acquisition and precipitation processing system is necessary to understand the source, quality and type of precipitation data available from the system. The NWS and other WSR-88D agency personnel interact with the radar system to provide warnings and other meteorologically related operational services. Access by others outside of the WSR-88D agencies is provided by an Internet-based system for dissemination.

The WSR-88D base data are produced by the Radar Data Acquisition (RDA) unit, which consists in a transmitter, receiver, signal processor, and an RDA computer. Base data from the RDA are referred to as Level II and consist in unprocessed (raw) reflectivity, Doppler wind velocity, and spectrum width. The next phase in the processing stream is the Radar Product Generator (RPG), which applies algorithms to produce the PPS products: 1-h accumulation, 3-h accumulation, storm total accumulation, and the hourly digital precipitation array (DPA). During the RPG processing, the data are aggregated in time and space, depending on the product generated. The precipitation products generally available outside of the NWS are aggregated at various resolutions and time intervals. From a hydrologic viewpoint, the space/time resolution and data precision of the generated products are important because these limitations directly affect hydrologic prediction when used as model input. The data and products generated by the WSR-88D system are characterized in terms of levels and stages. The term *Level* is used to describe single radar data and products:

- Level I is the analog signal coming from the receiver
- Level II is the base data containing raw reflectivity, spectral width and Doppler velocity
- Level III are derivative products generated from a single or dual polarized radar.

Dual polarization, in the vertical and horizontal planes, produces additional measures, called polarization diversity measurements, which are related to the rainfall rate and can outperform precipitation estimation based solely on single-polarized reflectivity (Doviak and Zrnic 1993). Ryzhkov et al. (2005a, b) documented significant improvements in rainfall estimation that can be achieved through the implementation of dual-polarization technology. These and other recently published comparisons compare precipitation estimates between horizontal-polarization Z-R estimates and multiparameter dual-polarization estimates of rainfall.

Products of particular interest to hydrology are precipitation products produced by the WSR-88D system. A principal dual-polarization product available from the WSR-88D radar is the digital precipitation rate (DPR). As processing algorithms are fine-tuned, DPR is expected to provide more accurate rainfall estimates, which is independent of rain gauges (Cocks et al. 2012). However, DPR accuracy can be variable as seen in Fig. 8.10 where the relative accuracy is expressed as an average difference with respect to rain gauges. All three products are in fairly close agreement between 12:00 and 15:00 UTC on May 23 2015, then after a hiatus, they are again close, except KEWX DPR begins to diverge significantly, while KGRK DPR remains consistent with KEWX GARR until 3:00 UTC on May 24.

The base data reflectivity recorded by a single radar is also of interest but requires additional processing for rainfall rate conversion from Level II reflectivity. As presented by Giangrande and Ryhkov (2008), recent dual polarization improvements demonstrate improvement of the dual polarization QPE (DPQPE) in comparison with rainfall estimates from the legacy NEXRAD PPS. From 612 gauge-radar pairs during 2012, accuracy was found to improve for the DPQPE compared with the PPS and with the greatest improvement for gauge totals exceeding 25.4 mm. Range dependent bias was less for the DPQPE indicating that it can achieve more reliable estimates at a far range from the radar. A subset of available data from the NEXRAD system is presented in Table 8.6 (NOAA-NWS

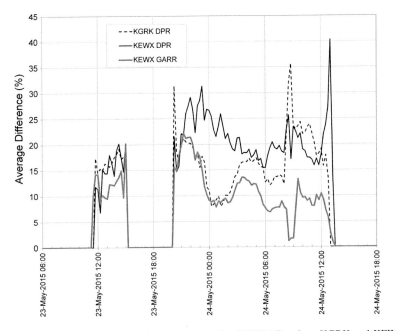

Fig. 8.10 Bias of DPR and GARR for two overlapping NEXRAD radars, KGRK and KEWX

Table 8.6 Level II base data and Level III products generated at a single radar site

Product	Range (km)	Precision	Resolution	Level
Base data—reflectivity	248	256-level	1 km × 1°	II
Base reflectivity	248	16-level	2 km × 1°	III
One-hour surface rainfall accumulation	124	16-level	2-km × 1°	III
Hourly digital precipitation array (DPA)	124	256-Level	4-km × 4-km	III
Digital instantaneous precipitation rate (DPR)	124	65536-Level	0.24 km × 1°	III

2015). The range column is the distance from the radar; precision indicates the number of data levels; resolution is shown in range and azimuth; and Level II or III indicates the location in the radar processing system where the data are derived.

These products are of principal interest to hydrologic modeling due to the high resolution in space and time and data precision. The DPA is a product of a moving window of 1-h duration and is georeferenced in the HRAP projection (Chap. 2). The DPR product shown on the last line of the table is in polar coordinates and derives its rainfall rates from dual polarization, independent of rain gauge bias correction.

8.4   Input for Hydrologic Modeling

Spatially distributed rainfall is important for accurate hydrologic prediction. Input to the distributed hydrologic model may consist of rain and/or snow. Radar, rain gauges, and possibly other sources of remotely sensed precipitation such as satellite can provide this source of input. In general, radar-based QPE must be bias corrected using rain gauges to remove systematic errors before input to a model. Other considerations are the space and time resolution of the rainfall measurements produced radar. The resolution of the input is dependent on the polar coordinate system used by the radar to measure reflectivity. Temporal resolution depends on the frequency that the volume of the atmosphere is scanned. The VCP of the radar defines the smallest temporal resolution of a given radar scan and has importance to the temporal rainfall resolution that a distributed model uses as input. The native resolution of the radar data can be resampled to coarser space and time scales. As mentioned, the sequence of VCPs followed by a radar during a storm causes time intervals to be uneven. Unless a hydrologic model is specifically designed to use rainfall data at uneven time increments, resampling is necessary.

Depending on where data are taken from the radar data processing stream, the data may have various resolution and data characteristics. Figure 8.11 shows a map of GARR rainfall produced over a 4,483 km^2 area in central Texas May 23–25, 2015. The data are derived from Level II reflectivity in polar coordinates and then sampled into a 1 × 1 km rectangular grid. In this case, post processing resulted in rainfall accumulations every 15 min and a spatial resolution that is 1 × 1 km in a rectangular coordinate system. An alternative sampling strategy is to group the radar measurements into subbasin averages for input to lumped models. The system called RainVieux that produced this rainfall data relies on real-time access to Level II reflectivity or Level III DPR (Vieux and Vieux 2003; and Looper and Vieux 2012).

Hydrologic prediction accuracy is particularly sensitive to radar rainfall bias as demonstrated by Vieux and Bedient (2004). The bias often introduced in radar results from assuming a Z-R relationship that is not representative of the actual drop size distribution during a particular storm event. Verification by an independent measurement, streamflow volume, revealed the importance, and effectiveness of bias correction of radar with rain gauge data. Controlling for bias in the radar input for five storms resulted in significant improvement in simulated runoff volume for a 260 km^2 urban catchment. Four events were tropical in nature during hurricane Frances and three periods during Tropical Storm Allison in 2001. A fifth event produced heavy rainfall generated by convective conditions on August 15, 2002. For the five events presented in Table 8.7, only one had a bias correction factor approximately equal to 1, which was Allison 1st with MFB = 0.971. This event relied on the tropical Z-R relationship, which was representative of actual raindrop distributions. Further, the average difference (AD) did not change markedly. Frances required a correction of MFB = 1.477. Introducing this correction, the Average Difference (AD) was halved from 32 to 16 %. The upstream areal-average

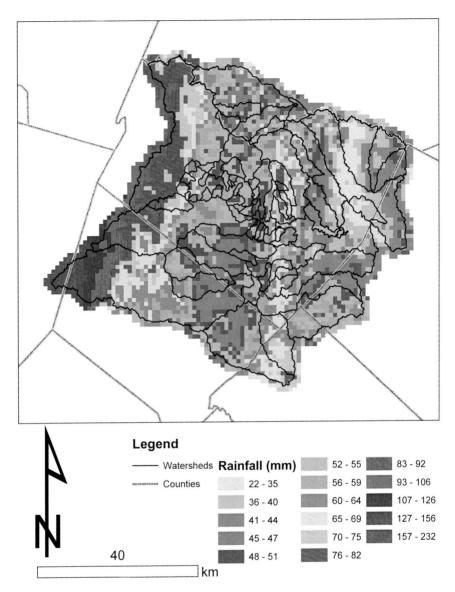

Fig. 8.11 Radar rainfall storm total in central Texas May 23–25, 2015

rainfall depth is seen in the last column and is the result of applying the bias correction. Whether this correction is important hydrologically is investigated by comparing the input depth with output streamflow. Note that this basin is heavily urbanized, composed of imperviousness exceeding 50 %. Otherwise soils are very clayey with an estimated saturated hydraulic conductivity of only 0.076 cm/hr.

Table 8.7 Event statistics showing improvement in rainfall accuracy resulting from mean field bias correction (Vieux and Bedient 2004)

Event	MFB	AD unadjusted (%)	AD adjusted (%)	Adjusted QPE depth (mm)	Unadjusted QPE depth (mm)
Frances	1.477	32.1	16.1	163.4	110.6
Allison 1st	0.971	15.9	15.9	39.6	40.8
Allison 2nd	0.623	68.3	21.2	74.3	119.3
Allison 3rd	0.636	61	19.0	96.7	152.0
August 15, 2002	0.836	23.8	11.7	101.1	120.9
Mean	0.909	40.2	16.8	95	108.7

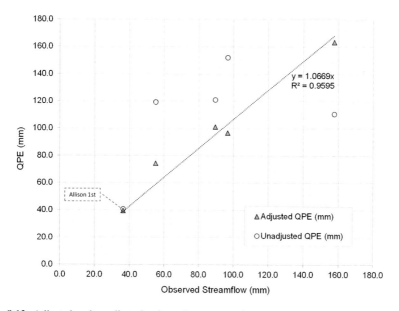

Fig. 8.12 Adjusted and unadjusted radar volume comparison to stream gauge volumes at Main Street

Given the imperviousness of the basins, most rainfall is converted to runoff and thus streamflow volume (output) can be used to validate radar-based QPE (input). Runoff volume is compared with two inputs: (1) QPE that is bias-corrected (second to last column in Table 8.7 above) and (2) QPE that relies on assumed Z-R relationships (last column in Table 8.7). Figure 8.12 shows the resulting comparison of QPE input volume versus streamflow output at the stream gauge, Main Street (see Vieux and Bedient 2004). The near 1:1 trend line fitted to the adjusted QPE versus observed streamflow has a coefficient of determination, $R^2 = 0.9595$, whereas the unadjusted QPE only has an $R^2 = 0.2327$. While the Allison 1st storm event did not require a bias adjustment (MFB = 0.971) as indicated, the other events with MFB significantly different from 1 shows considerable divergence when their volume is plotted versus observed streamflow. Looper et al. (2012) found significant improvement in modeled streamflow accuracy when radar-based rainfall estimates were reprocessed through additional quality control of rain gauge and radar bias correction of NWS Multisensor Precipitation Estimates (MPE). Using this gauge-corrected precipitation estimates (GCPE), a 74 % improvement resulted in the Nash Sutcliffe statistical performance of a hydrologic model of the Blue River and Illinois River, both located in Oklahoma. One explanation for bias in the NWS MPE is that available rain gauges are not adequately quality controlled, or perhaps even used in operational QPE production. The NWS does not maintain records of which rain gauges were used or excluded in their QPE production process. These results confirm that bias adjustment of radar-derived QPE has hydrologic significance beyond just making radar-based QPE agree with rain gauges.

8.5   Summary

Distributed rainfall containing high-resolution spatially and temporally variable rainfall is an important component of hydrologic modeling. The required numbers of gauges used to characterize the rainfall field depends on local variability induced by terrain or coastal gradients. Increasing the number of closely spaced rain gauges reduces the standard error of the mean up to a certain amount dictated by point process variability and spatial autocorrelation. There are progressively fewer benefits as gauges are added to the network due to correlation that reduces the independence of the added gauge. Rainfall derived from radar has advantages of producing high-resolution rainfall over broad areas at time intervals around 5 min and resolution of approximately 1 km in range and 1 degree in azimuth. However, dense the radar QPE is, it likely contains bias that diminishes its utility in hydrologic modeling. Overcoming bias is achieved by comparing rain gauge and radar accumulations and making corrections to remove systematic error. Beyond just making the radar agree with rain gauges, bias correction of radar QPE has hydrologic significance as demonstrated by independent streamflow measurements for a nearly impervious basin. While rain gauge networks are used to improve on the assumed relationship between radar reflectivity and rainfall rate, the Z-R relationship, advances in dual polarization shows promise, especially because it does not rely on rain gauges and attendant gauge errors. Combined use of radar and gauge networks produces more accurate precipitation measurements. Considering the importance of rainfall input to hydrologic models, the distributed nature of radar rainfall is proving to be a significant advance in hydrologic modeling.

References

Ambroise, B., and J. Aduizian-Gerard. 1989. Test of a trigonometrical model of a slope rainfall in a small rengelbach catchment. In *Proc WMO/IAHS/ETH Workshop*, ed. Sevruk, B. Swiss Federal Institute of Technology, Zurich, 81–85. St. Moritz. Switzerland, 4–7 December 1989.

Bedient, P.B., A. Holder, J.A. Benavides, and B.E. Vieux. 2003. Radar-based flood warning system applied to tropical storm allison. *Journal of Hydrologic Engineering* 8(6): 308–318.

Berkowitz, D.S., J.A. Schultz, S. Vasiloff, K.L. Elmore, C.D. Payne, and J.B. Boettcher. 2013. Status of Dual Pol QPE in the WSR-88D Network. In: *93th AMS, 27th conference on Hydrology*. Austin, TX.

Chrisman, J., and C. Chrisman. 1999. An operational guide to WSR-88D reflectivity data quality assurance. In: *WSR-88D Radar Operations Center paper*, 15 pp. (Available from WSR-88D Radar Operations Center, 3200 Marshall Ave., Norman, OK 73072.).

Cocks, S.B., D.S. Berkowitz, R. Murnan, J.A. Schultz, S. Castleberry, K. Howard, K. Elmore, and S. Vasiloff, 2012. Initial assessment of the dual-polarization quantitative precipitation estimate algorithm's performance for two dual-polarization WSR-88Ds. Proceedings of 28th conference on interactive information processing systems (IIPS), New Orleans, LA, Amer. Meteor. Soc., 7B.2.

Crum, T.D., and R.L. Alberty. 1993. The WSR-88D and the WSR-88D operational support facility. *Bulletin of the American Meteorological Society* 27(9): 1669–1687.

Doviak, R.J., and D.S. Zrnic. 1993. *Doppler radar and weather observations*, 2nd ed. Orlando, Florida: Academic Press.

Einfalt, T., K. Arnbjerg-Nielsen, D. Faure, N.-E. Jensen, M. Quirmbach, G. Vaes, B.E. Vieux, and C. Golz. 2004. Towards a roadmap for use of radar rainfall data in urban drainage. *Journal of Hydrology* 299(3–4): 186–202.

Emmanuel, I., H. Andrieu, E. Leblois, and B. Flahaut. 2012. Temporal and spatial variability of rainfall at urban hydrological scales. *Journal of Hydrology* 430–431: 162–172.

Emmanuel, I., H. Andrieu, E. Leblois, N. Janey, and O. Payrastre, 2015. Influence of rainfall spatial variability on rainfall–runoff modelling: benefit of a simulation approach? *Journal of Hydrology* 531, Part 2: 337–348.

Faurès, J.-M., D.C. Goodrich, D.A. Woolhiser, and S. Sorooshian, 1995. Impact of small-scale spatial rainfall variability on runoff modeling. *Journal of Hydrology* 173(1–4): 309–326.

Fulton, R.A., J.P. Breidenbach, D.-J. Seo, D.A. Miller, and O'Bannon, T. 1998. The WSR-88D Rainfall Algorithm. *Journal of Weather and Forecast* 13(2): 377–395.

Giangrande, S.E., and A.V. Ryzhkov. 2008. Estimation of rainfall based on the results of polarimetric echo classification. *Journal of Applied Meteorology and Climatology* 47: 2445–2462.

Goodrich, D.C. 1990. Geometric simplification of a distributed rainfall-runoff model over a range of basin scales. PhD. diss., University of Arizona, Tucson, AZ.

Goodrich, D.C., J.-M. Faurès, D.A. Woolhiser, L.J. Lane, and S. Sorooshian, 1995. Measurement and analysis of small-scale convective storm rainfall variability. *Journal of Hydrology* 173(1–4): 283–308.

Goutorbe, J.-P., T. Lebel, A. Tinga, P. Bessemoulin, J. Bouwer, A.J. Dolman, E.T. Wingman, J.H. C. Gash, M. Hoepffner, P. Kabat, Y.H. Kerr, B. Monteny, S.D. Prince, F. Saïd, P. Sellers, and J.S. Wallace. 1994. Hapex-sahel: a large scale study of land-surface interactions in the semi-arid tropics. *Annales Geophysicae* 12(1): 53–64.

Haan, C.T. 2002. *Statistical methods in hydrology*. Ames, Iowa: Iowa State University Press. ISBN 978-0813815039.

Lebel, T., and L. Le Barbé. 1997. Rainfall monitoring during HAPEX-Sahel. 2. Point and areal estimation at the event and seasonal scales. *Journal of Hydrogeology* 188–189: 97–122.

Lebel, T., J.D. Taupin, and N.D'Amato, 1997. Rainfall monitoring during HAPEX-Sahel. 1. General rainfall conditions and climatology. *Journal of Hydrogeology* 188–189: 74–96.

Lebel, T., B. Cappelaere, S. Galle, N. Hanan, L. Kergoat, S. Levis, B.E. Vieux, L. Descroix, M. Gosset, E. Mougin, C. Peugeot, and L. Seguis, 2009. AMMA-CATCH studies in the Sahelian region of West-Africa: an overview. *Journal of Hydrogeology* 375(1–2): 3–13.

Looper, J.P., and B.E. Vieux. 2012. An assessment of distributed flash flood forecasting accuracy using radar and rain gauge input for a physics-based distributed hydrologic model. *Journal of Hydrogeology* 412: 114–132.

Looper, J.P., B.E. Vieux, and M.A. Moreno, 2012. Assessing the impacts of precipitation bias on distributed hydrologic model calibration and prediction accuracy. *Journal of Hydrogeology* 418–419: 110–122.

Marshall, J.S. and W. Mc K. Palmer, 1948 The Distribution of Raindrops with Size. *Journal of meteorology* 5: 165–166.

Mimikou, M.A., and E.A. Baltas, 1996. Flood forecasting based on radar rainfall measurements. *Journal of Water Resources Planning and Management* 122(3).

Morin, E., D.C. Goodrich, R.A. Maddox, X. Gao, H.V. Gupta, and S. Sorooshian, 2006. Spatial patterns in thunderstorm rainfall events and their coupling with watershed hydrological response. *Advances in Water Resources* 29(6): 843–860. ISSN 0309-1708.

Morin, J., D. Rosenfield, and E. Amitai. 1995. Radar rain field evaluation and possible use of its high temporal and spatial resolution for hydrological purposes. *Journal of Hydrogeology* 172: 275–292.

NOAA-NWS, 2015. Interface Control Document, Build 16.0, Document 26200001 V. WSR-88D Radar Operations Center. Norman, OK.

Pathak, C. and B. Vieux, 2008. *Geo-spatial comparison of rain gauge and nexrad data for Central and South Florida.* World Environmental and Water Resources Congress, May, 1–11.

Pereira, A.J., K.C. Crawford, C.L. Hartzell., 1998. Improving WSR-88D hourly rainfall estimates. *Journal of Weather and Forecasting, American Meteorological Society,* 13: 1016–1028.

Ryzhkov, A.V., S.E. Giangrande, and T.J. Schuur. 2005a. Rainfall estimation with a polarimetric prototype of the WSR-88D Radar. *Journal of Applied Meteorology* 44: 502–515.

Ryzhkov, A.V., T.J. Schuur, D.W. Burgess, P.L. Heinselman, S. Giangrande, and D.S. Zrnic. 2005b. The joint polarization experiment: polarimetric rainfall measurements and hydrometeor classification. *Bulletin of the American Meteorological Society* 86: 809–824.

Rosenfeld, D., D.B. Wolff, and D. Atlas. 1993. General probability-matched relations between radar reflectivity and rain rate. *Journal of applied Meteorology* 32: 50–72.

Rosenfeld, D., D.B. Wolff, and E. Amitai. 1994. The window probability method for rainfall measurements with radar. *Journal of Applied Meteorology* 33: 682–693.

Rosenfeld, D., E. Amitai, and D.B. Wolff. 1995a. Classification of rain regimes by the 3-dimensional properties of reflectivity fields. *Journal of Applied Meteorology* 34: 198–211.

Rosenfeld, D., E. Amitai, and D.B. Wolff. 1995b. Improved accuracy of radar WPMM estimated rainfall upon application of objective classification criteria. *Journal of Applied Meteorology* 34: 212–223.

Seo, D.-J., J.P. Breidenbach, and E.R. Johnson. 1999. Real-time estimation of mean field bias in radar rainfall data. *Journal of Hydrology* 223: 131–147.

Seo, D.-J., and P. Breidenbach. 2002. Real-time correction of spatially nonuniform bias in radar rainfall data using rain gages measurements. *Journal of Hydrometeorology* 3: 93–111.

U.S. Army Corps of Engineers, 1994. *Flood runoff analysis.* Engineer Manual 1110–2-1417, Washington, DC.

U.S. Army Corps of Engineers, 1996. *Hydrologic aspects of flood warning—preparedness programs.* Technical Letter 1110-2-540, Washington, DC.

U.S. Department of Commerce. 1947. *Thunderstorm rainfall,* Hydrometeorological Report No. 5, Weather Bureau, Office of Hydrologic Director, Silver Springs, MD.

U.S. Department of Commerce, 2013. Federal Meteorological Handbook No. 11, Doppler Radar Meteorological Observations: Part A System Concepts, Responsibilities, and Procedures, FMC-H11A-2003, Washington, DC.

U.S. Department of Commerce, 2005. Federal meteorological handbook no. 11, *Doppler radar meteorological observations: part B doppler radar theory and meteorology*, FCM-H11B-2005, Washington, DC.

U.S. Department of Commerce, 2006a. Federal meteorological handbook no. 11, *Doppler radar meteorological observations: part C WSR-88D products and algorithms*, FCM-H11C-2006, Washington, DC.

U.S. Department of Commerce, 2006b. Federal meteorological handbook no. 11, *Doppler radar meteorological observations: part D WSR-88D unit description and operational applications*, FCM-H11D-2006, Washington, DC.

Vallabhaneni, S., B.E. Vieux, and T. Meeneghan, 2003. Radar-rainfall technology integration into hydrologic and hydraulic modeling projects. In *Practical Modeling of urban water systems, monograph 12*. Proceedings of the 2003, Stormwater and Urban Water Systems Modeling Workshops and Conference, Toronto Canada. Computational Hydraulics Institute.

Vieux, B.E., and P.B. Bedient. 1998. Estimation of rainfall for flood prediction from WSR-88D reflectivity: a case study, 17-18 October 1994. *J. Weather and Forecast* 13(2): 407–415.

Vieux, B.E., and P.B. Bedient, 2004. Assessing urban hydrologic prediction accuracy through event reconstruction. *Journal of Hydrogeology*, Special Issue on Urban Hydrology. Forthcoming.

Vieux, B.E. and J.E. Vieux, 2003. Development of a radar rainfall system for sewer system management. In *Proceedings of sixth international workshop on precipitation in Urban areas measured and simulated precipitation data requirements for hydrological modelling*, 4–7 December, Pontresina, Switzerland.

Vieux, B. and Pathak, C., 2007. Evaluation of rain gauge network density and NEXRAD rainfall accuracy. In *Proceedings of the American Society of civil engineers*, World Environmental and Water Resources Congress, 1–12. doi:10.1061/40927(243)278.

Wilson, J., and E. Brandes. 1979. Radar measurement of rainfall—a summary. *Bulletin of the American Meteorological Society* 60: 1048–1058.

Zawadzki, I.I. 1973. Statistical properties of precipitation patterns. *Journal of Applied Meteorology* 12: 459–472.

Zawadzki, I.I. 1975. On radar-raingage comparison. *Journal of Applied Meteorology* 14: 1430–1436.

Chapter 9
Surface Runoff Model Formulation

Abstract Hydrologic and environmental processes are distributed in space and time. Simulation of these processes is made possible through the already well-developed spatial data analysis and management techniques of a GIS. Digital maps of soils, land use, topography, and rainfall are used to compute rainfall runoff in each grid cell in the drainage network. In principle, runoff generation caused by rainfall rates exceeding infiltration rates or soil profile saturation can is simulated in this scheme. Runoff losses due to infiltration in channels can account for runoff processes typical of alluvial fans in more arid climates or due to karstic geology where fractures permit runoff arriving from upstream to percolate into the subsurface or aquifer as recharge. The objective of this chapter is to explore the model formulation and the geospatial data used to define topography, land use/cover, soils, and precipitation input within a physics-based distributed framework.

9.1   Introduction

The digital watershed shown in Fig. 9.1 is composed of finite elements that connect each grid cell together according to the principal drainage direction. The connectivity of the finite elements forms the basis for solving the kinematic wave equations. The conservation equations are used to model explicitly the hydraulic components of the drainage network. Overland flow, channel hydraulics, storage–discharge relationships for detention basins, complex channel cross sections, stage-discharge rating curves, and shallow water wave propagation are combined where appropriate to model the digital watershed. The close coupling of runoff generation and losses and the routing of this runoff through the drainage network are achieved through the *hydraulic* approach to *hydrology*. This approach is a departure from traditional methods such as the unit hydrograph where runoff generation and routing are artificially separated. This chapter presents the mathematical analogy and numerical algorithms that provide the foundation for physics-based distributed hydrologic modeling using conservation of mass and momentum.

Analytic solutions to the equations governing runoff are not generally obtainable giving rise to the need for numerical methods such as finite element or finite difference

© Springer Science+Business Media Dordrecht 2016   165
B.E. Vieux, *Distributed Hydrologic Modeling Using GIS*,
Water Science and Technology Library 74, DOI 10.1007/978-94-024-0930-7_9

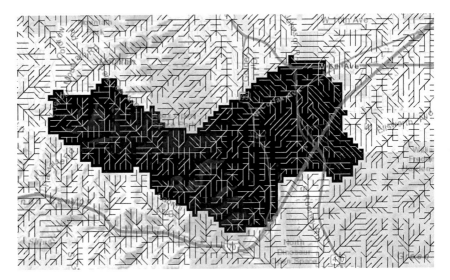

Fig. 9.1 Grid-cell representation and drainage network used in the finite element method to simulate watershed runoff using V*flo*®. Connectivity of the elements is derived from the digital elevation map of flow direction

methods (Singh and Woolhiser 2002). Jayawardena and White (1977, 1979), Ross et al. (1979), and Kuchment et al. (1983, 1986), among others, used the finite element method to solve the governing equations on equivalent cascades, planes, or subareas with homogeneous properties. The solution using linear, one-dimensional elements presented by Vieux et al. (1988, 1990) used a single chain of finite elements for solving overland flow. This solution differed from previous finite element solutions because it represented roughness and slope as nodal rather than elemental parameters. This difference in approach enables the simulation of spatially variable watershed surfaces without the need to break the watershed into equivalent conceptual cascades, planes, or subareas. The PBD model called *r.water.fea* is a finite element approach to watershed modeling that explicitly represents spatially variable input and parameters. This model was developed in 1993 for the U.S. Army Corps of Engineers, Construction Engineering Research Laboratory, Champaign, Illinois (CERL). The goal was to provide a hydrologic modeling tool that used GIS maps of parameters directly within the GIS environment. The initial development of the model is a part of the public domain GIS called GRASS (Geographic Resource Analysis Support System). Vieux and Gauer (1994) extended this finite element solution to a network of elements representing a watershed domain with channels within a GIS environment. Later modifications added channel routing, a Green and Ampt infiltration routine and distributed radar rainfall input. Finite elements are laid out in the direction of the principal land surface slope, which is consistent with the kinematic flow analogy. Unlike other finite element applications where high element density is used to resolve large gradients or complex geometry, the elements used by *r.water.fea* are one-dimensional elements with lengths consistent with the grid-cell resolution. This

constant resolution supports the parameterization of the nodal finite element parameters using raster maps with grid-cell values. Through linkage of the raster data
structure and the nodal finite element method, the grid-cell value becomes the nodal
value in the finite element solution.

Distributed hydrologic modeling may be termed physics-based if it uses conservation of momentum, mass, and energy to model the processes. Most
physics-based models solve a flow analogy (e.g., kinematic, diffusive wave, or full
dynamic) with numerical methods using a discrete representation of the catchment,
such as a finite difference grid or finite element mesh. Besides *r.water.fea* described
in Vieux and Gauer (1994) and Vieux (2001), other PBD models include the V*flo*$^{®}$
distributed hydrologic model (Vieux and Vieux, 2002; Vieux et al. 2003); CASC2D
(Julien and Saghafian, 1991; Ogden and Julien 1994); GSSHA gridded diffusive
wave model (Downer and Ogden, 2004); Systeme Hydrologique Européen
(SHE) (Abbott et al. 1986a, b); Distributed Hydrology Soil Vegetation Model
(DHSVM) (Wigmosta et al. 1994); DEM-based overland flow routing model of
surface runoff using diffusive wave analogy (Jain and Singh 2005); Kinematic
Wave Method for Subsurface and Surface runoff (KWMSS) (Lee et al. 2009);
gridded conceptual model called HL-RDHM, which is a modified version of
SAC-SMA with kinematic hillslope and channel routing (Koren et al. 2004, 2012);
complex process modeling that integrates land surface modeling with surface and
subsurface runoff developed within the WRF-Hydro framework (Gochis et al. 2014;
Clark et al. 2015a, b); and Triangular Irregular Network (TIN)-based model with
surface and subsurface runoff coupling (Ivanov et al. 2004a, b; Vivoni et al. 2004
and 2007). All of the above-cited models employ some form of digital terrain
representation as discrete elements, either as grids or TINs.

The *r.water.fea* and V*flo*$^{®}$ models employ the finite element numerical solution
of elements connecting grid cells. Many of these models are either internally
integrated with a GIS or setup externally with required parameters derived from
geospatial data and GIS analysis. The model *r.water.fea* was integrated as an
internal module of GRASS as a GIS function to compute runoff volume and peak.
While V*flo*$^{®}$ does use geospatial data, it *does not* require a GIS to run the model.
Digital terrain processing for creation of the drainage network and assignment of
slope at each node is internal to V*flo*$^{®}$ version 6.x (User Manual 2016). Integration
of V*flo*$^{®}$ in ArcGIS resulted in the drainage network shown in Fig. 9.2. GIS integration facilitates the analysis of geospatial data, extraction of the drainage network,
and creation of distributed parameter maps. Besides drainage direction, nodal
parameters in the finite element solution are taken from the raster of gridded values
such as hydraulic roughness. The GIS raster data become the parameters in the
model making an integrated model. GIS integration was used for developing basin
models and for coupling a structured mesh model, V*flo*$^{®}$, with an unstructured mesh
used in the shallow water model, ADCIRC (Tromble et al. 2010).

Besides being an efficient means for solving the KWA equations, the finite
element method provides an intuitive approach where arrows (1D finite elements)

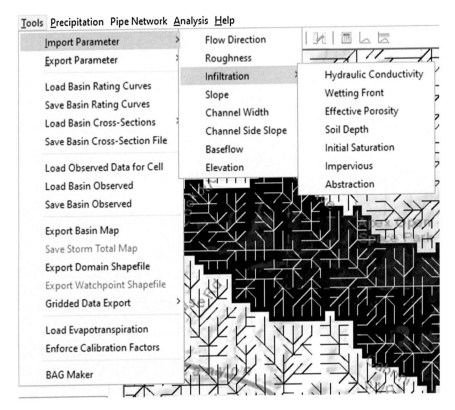

Fig. 9.2 V*flo*® map analysis used in the creation of parameter maps

are laid out between grid cells in the direction of principal slope. Taking a kernel composed of a 3 × 3 cell patch, the drainage network and computational scheme may be visualized. Figure 9.3 shows the grid-cell scheme used to define the finite elements connecting together overland flow and channel elements. The connectivity of the drainage network is used to develop the system of equations that provide a solution to the kinematic wave analogy. The drainage network itself and the nodal slope values are derived from a DEM (Fig. 9.4).

When the diagonals are considered, flow direction is referred to as D8. Thus, there may be up to eight flow directions that converge on a single cell. Flat or divergent flow areas where the flow direction is indeterminate must be resolved before such cells may be incorporated into the solution. The resolution of the DEM determines the fundamental length scale for routing water over the land surface and through channels. How the D8 directions and the drainage network are used to solve the mathematical flow analogy is described below.

Fig. 9.3 Schematic representation of runoff in a grid-based finite element model solution

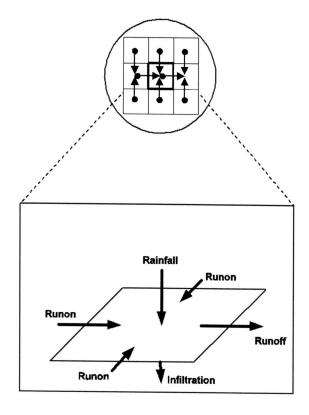

Fig. 9.4 Eight drainage directions defining connectivity for each cell in the DEM

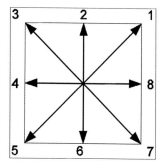

9.2 Mathematical Formulation

The St. Venant equations *full dynamic* momentum mathematical analogies are simplifications where gradients of lesser magnitude are ignored, as shown in Eq. 9.1. There are two simplifications, the diffusive wave and the kinematic wave analogy (KWA) velocity, V, and the flow depth, y, in a prismatic channel, defined as follows:

$$S_f - S_o = 0 \qquad\qquad \text{Kinematic}$$
$$\frac{\partial y}{\partial x} + S_f - S_o = 0 \qquad\qquad \text{Diffusion} \qquad\qquad (9.1)$$
$$\frac{\partial V}{\partial t} + V\frac{\partial V}{\partial t} + g\left(\frac{\partial y}{\partial x} + S_f - S_o\right) = 0 \quad \text{Dynamic}$$

where S_f and S_o are the friction and bed slope gradients, respectively. KWA equations assume that the friction gradient is equivalent to the bottom slope. More terms are included in the diffusion analogy and dynamic forms of the equation. If backwater or flat slopes are prevalent, the DWA or full dynamic equations may be more appropriate to the solution analogy. Lighthill and Whitham (1950) explored kinematic wave motion used in describing flood waves in rivers. If all other terms are small, or an order of magnitude less than the bed slope or friction gradient, the KWA is an appropriate representation of the wave movement downstream (Chow et al. 1988).

The KWA consists in two equations, the conservation of mass and momentum relationship. The one-dimensional continuity equation for overland flow resulting from rainfall excess is expressed by

$$\frac{\partial h}{\partial t} + \frac{\partial (uh)}{\partial x} = R - I \qquad\qquad (9.2)$$

where R is the rainfall rate; I is the infiltration rate; h is the flow depth; and u is the overland flow velocity. The forcing term on the right is rainfall excess intensity, which drives the hydrologic response, often referred to as *rain on grid* method. By definition, when the friction gradient is assumed equal to the channel bed slope, the result is the *uniform flow* assumption. An appropriate relation between velocity, u, and flow depth, h, such as the Manning equation is

$$u = \frac{S_o^{1/2}}{n} h^{2/3} \qquad\qquad (9.3)$$

where S_o is the bed slope or principal land surface slope and n is the hydraulic roughness. In the KWA, velocity and flow depth depend on the land surface slope. For this reason, the KWA enforces that

1. The slope of the water surface and friction gradient are parallel with the land surface slope
2. The flow is *uniform*
3. Backwater is not admitted in the solution

Substituting Eq. 9.3 into Eq. 9.2 results in the KWA equation for relating flow depth to the land surface slope, hydraulic roughness, rainfall, and infiltration rates:

$$\frac{\partial h}{\partial t} + \frac{S_o^{1/2}}{n}\frac{\partial h^{5/3}}{\partial x} = R - I \qquad\qquad (9.4)$$

For channelized flow, Eq. 9.4 is written in terms of the cross-sectional area A instead of the flow depth h:

$$\frac{\partial A}{\partial t} + \frac{\partial Q}{\partial x} = q \qquad (9.5)$$

where Q is the discharge or flow rate in the channel and q is the rate of lateral inflow per unit length in the channel. Equations 9.4 and 9.5 are the governing equations for the KWA applied to overland and channel flow. Next, we develop numerical solutions of the KWA for a grid-cell representation of the watershed.

9.2.1 Numerical Solution

The finite element method transforms partial differential equations in *space and time* into ordinary differential equations in *time*. Many different types of finite element solutions have been developed in the general field of engineering mechanics (Segerlind 1984). Galerkin's formulation has been used to solve Eqs. 9.4 and 9.5 using 1D linear elements (Meadows and Blandford 1990; Vieux 1988; Vieux and Segerlind 1989; Vieux et al. 1990; Kazezyılmaz-Alhan and Medina 2007). The dependent variables, A and Q, are written in terms of elemental shape functions as

$$A^{(e)} = \sum_{i=1}^{N} N_i(x)A_i$$
$$\qquad (9.6)$$
$$Q^{(e)} = \sum_{i=1}^{N} N_i(x)Q_i$$

where $Q^{(e)}$ and $A^{(e)}$ are the approximating functions for the discharge and the cross-sectional area, respectively. The superscript, (e), indicates that the functions are applied to individual elements. Interpolation within the finite element method relies on $N_i(x)$, which is a linear shape function. For more details on the finite element method, the reader is directed to the many finite element books, such as Segerlind (1984) who gives a clear presentation on the finite element method.

In Galerkin's formulation, the weighting and shape functions are of the same form. The system of equations formed by these approximating functions called the residual is minimized when integrated over the entire problem domain, Ω. Thus the residual for the 1D conservation of mass equation for channel flow is

$$R^{(e)} = \int_{\Omega} N^T \left[\frac{\partial A}{\partial t} + \frac{\partial Q}{\partial x} - q \right] = 0 \qquad (9.7)$$

For overland flow and $q = u \cdot h$, the residual becomes

$$R^{(e)} = \int_{\Omega} N^T \left[\frac{\partial h}{\partial t} + \frac{\partial q}{\partial x} - (R - I) \right] = 0 \qquad (9.8)$$

Substituting shape function approximations to $A^{(e)}$ and $Q^{(e)}$ in Eq. 9.6, the elemental residual is obtained, leading to the *consistent* formulation for channel flow:

$$R^{(e)} = \frac{L}{6}\begin{bmatrix} 2 & 1 \\ 1 & 2 \end{bmatrix}\dot{A} + \frac{1}{2}\begin{bmatrix} -1 & 1 \\ -1 & 1 \end{bmatrix}Q - \frac{qL}{2}\begin{bmatrix} 1 \\ 1 \end{bmatrix} = 0 \qquad (9.9)$$

or in matrix form:

$$C\dot{A} + BQ = F \qquad (9.10)$$

where A is the derivative with respect to time. C and B are matrices and A, Q, and F are vectors of dependent variables, cross-sectional area, flow and rainfall excess forcing, respectively. The product Q with B represents the gradients in space of flow rate or $\partial Q/\partial x$. The formulation expressed by Eq. 9.9 is often referred to as the *consistent* formulation, because the linear variation of the function $\partial A/\partial t$ with respect to x is consistent with the linear variation assumed for $A(x)$.

Out of concern for efficient memory use, as well as computational time, the lumped formulation can be applied to Eq. 9.9. Lumping is a feature of the finite element method and in this context merely refers to diagonalizing the capacitance matrix, C. It assumes that $\partial A/\partial t$ with respect to x is constant between the midpoints of adjacent elements, producing a diagonal capacitance matrix. The lumped form for channel flow, Q, is

$$R^{(e)} = L\begin{bmatrix} 1 & 0 \\ 0 & 1 \end{bmatrix}\dot{A} + \begin{bmatrix} -1 & 1 \\ -1 & 1 \end{bmatrix}Q - qL\begin{bmatrix} 1 \\ 1 \end{bmatrix} = 0 \qquad (9.11)$$

For overland flow, the lumped form becomes

$$R^{(e)} = L\begin{bmatrix} 1 & 0 \\ 0 & 1 \end{bmatrix}\dot{h} + \begin{bmatrix} -1 & 1 \\ -1 & 1 \end{bmatrix}q - (R-I)L\begin{bmatrix} 1 \\ 1 \end{bmatrix} = 0 \qquad (9.12)$$

Equations 9.11 and 9.12 are suitable for solving a single chain of linear elements. Simulating hillslopes with this formulation is appropriate, because it treats the flow as having a single gradient, $\partial h/\partial x$, in the x-direction measured by L. Spatially variable parameters of slope, infiltration, and hydraulic roughness are represented in the direction of flow but not transversally. There are no transverse gradients, say, in the y-direction, because this term is not accounted for in the conservation of mass equation. This approach has several advantages. First, only one gradient $\partial Q/\partial x$, which is the direction of the principal land surface gradient, needs to be computed. Second, as we will see, the analogy is consistent with the GIS representation of drainage direction using grid cells. Finite solution of two-dimensional domains using 1D finite elements is termed a *partial discretization*. An advantage of this approach is that it saves computational effort for domains where the KWA is valid.

9.2.2   Grid Resolution Effects

To extend the formulation to simulate surface flow where a number of branches form a network draining the watershed area, we must modify the form of Eqs. 9.11 and 9.12. This can be done by taking the average of summed inflow to a cell. We use the same idea of the elemental residual, as in Eqs. 9.7 and 9.8 but modify it to account for more than one element arriving at the same node; we obtain Eq. 9.13 for a network of elements. Since we are now introducing a width w_{eq} to account for the watershed area but using 1D finite elements, we rewrite Eq. 9.11 for overland flow through a drainage network:

$$R^{(e)} = L \begin{bmatrix} 0 & 0 \\ 0 & 1 \end{bmatrix} \dot{A} + \begin{bmatrix} 1/\kappa_i & 0 \\ -1 & 1/\kappa_j \end{bmatrix} Q - (R - I) w_{eq} L \begin{bmatrix} 0 \\ 1 \end{bmatrix} = 0 \qquad (9.13)$$

where κ_i and κ_j are the valences at the ith (upstream) and jth (downstream) node, respectively; $R^{(e)}$ and w_{eq} are the elemental width such that the entire drainage area is correctly represented using one-dimensional elements. Thus, the elemental width is an equivalent width that is simply the total drainage area divided by the total drainage length. The equivalent width, w_{eq}, is computed as

$$w_{eq} = \frac{\text{Area}_{total} - \text{Area}_{chan}}{\text{Length}_{total} - \text{Length}_{chan}} \qquad (9.14)$$

Excluding the channel area and length in Eq. 9.14 is necessary for watersheds where the channel area is a large fraction of the total area. In most general watershed applications, this modification is not necessary. This formulation is valid as long as the watershed flow elements lead to just one outlet and there is no bifurcation in the downstream direction at any cell.

9.3   Surface Runoff Modeling Example

In this section, a steady-state example illustrates how to form the stiffness matrix and the forcing vector using the equations previously demonstrated. Finding a physically and computationally realistic resolution is an important task that influences model calibration and performance. The resolution chosen affects both the drainage network length and the size of the system of equations assembled using the finite element method. Physically, we should consider how realistically the GIS data base represents the following:

- drainage network
- spatial variability of parameters
- precipitation input.

Fig. 9.5 Representation of
the 3 × 3 kernel and finite
element connectivity

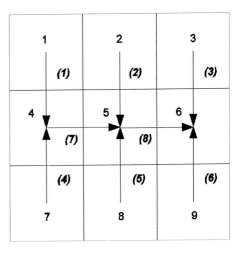

The D8 flow direction map provides the principal slope direction and defines the layout of finite elements forming the drainage network. The connectivity of these elements is used to assemble the finite elements into a system of equations as demonstrated below. For an example of watershed, the connectivity is shown in Fig. 9.5.

The element numbers are shown as italic numbers in parentheses. The node numbers are adjacent to the node of each finite element. This also corresponds to the center of the grid cell in the GIS database where each parameter value resides. The direction of each element, from upstream to downstream, is derived from the D8 flow direction map. In this example, no diagonal drainage directions are present to simplify computations, whereas in the model implementation, diagonal cells are considered.

The process for developing the connectivity of a drainage network for finite element modeling is as follows:

- Number each grid cell with a unique number.
- Layout finite elements according to the drainage direction.
- Identify the grid-cell number at the upstream and downstream nodes.
- Form the connectivity table.
- Number of elements meeting at each node (in- and outbound) becomes the valence of the node.
- Identify boundary condition nodes on watershed divides and assign zero flow depth for all time.

Following this process for the configuration shown above in Fig. 9.5, we obtain the results as in Table 9.1. The last two columns on the right are the valences or the number of elements that meet at each node.

From a numerical solution viewpoint, the boundary conditions are the cells where the flow depth and velocity are held equal to zero for all time. From the viewpoint of the watershed representation using grid cells, this corresponds to cells

Table 9.1 Finite element connectivity for 3 × 3 grid

Element	Upstream (from)	Downstream (to)	κ_i	κ_j
1	1	4	1	3
2	2	5	1	4
3	3	6	1	3
4	7	4	1	3
5	8	5	1	4
6	9	6	1	3
7	4	5	3	4
8	5	6	4	3

on ridgelines both along the watershed boundary and within the interior of the watershed. Whenever the valence is equal to one, a boundary condition is imposed.

Taking each element, we can assemble a system of equations using the elemental residual Eq. 9.13. To see how this is performed, we can take one element and add the nodal contributions of each element to the global system. Taking the first element and identifying the upstream and downstream nodes from Table 9.1, we obtain the elemental residual for Element 1:

$$R^{(e)} = \frac{1}{4}L\begin{bmatrix} 0 & 0 \\ 0 & 1 \end{bmatrix}\dot{A} + \begin{bmatrix} 1/\kappa_i & 0 \\ -1 & 1/\kappa_j \end{bmatrix}Q - (R-I)w_{eq}L\begin{bmatrix} 0 \\ 1 \end{bmatrix} = 0 \qquad (9.15)$$

where the numbers show the upstream node 1 and downstream node 4, relating the nodes to the global system of equations. Assembly of the elemental residuals into the global system of equations for the entire watershed is performed according to the well-known direct stiffness method (Segerlind 1984).

Using the nomenclature of (*row, col*) referring to the location in the global matrix, we add the nodal values in Eq. 9.15 to the stiffness matrix B in Eq. 9.16 at the respective locations. We have a value of 1/1 at (1,1); 0 at (1,4); −1 at (4,1); and 1/3 at (4,4). Similarly, matrix C and forcing vector F may be written as

$$
\begin{matrix} 1 \\ 2 \\ 3 \\ 4 \\ 5 \\ 6 \\ 7 \\ 8 \\ 9 \end{matrix}
\begin{pmatrix} 1 & & 0 & & \\ & & & & \\ & & & & \\ -1 & & 1/3 & & \\ & & & & \\ & & & & \\ & & & & \\ & & & & \\ & & & & \end{pmatrix}
\quad Q = w_{eq}(R-I)
\begin{pmatrix} 0 \\ 0 \\ 0 \\ L^{(1)} \\ 0 \\ 0 \\ 0 \\ 0 \\ 0 \end{pmatrix}
\qquad (9.16)
$$

At this stage, only the length of element number one, $L^{(1)}$, is added to the forcing vector, F. Once we have cycled through each element, adding its contributions and applied boundary conditions, we obtain the following for the steady-state case in Eq. 9.17:

$$
\begin{pmatrix}
1 & 0 & & & & & & & \\
0 & 1 & 0 & & & & & & \\
& 0 & 1 & 0 & & & & & \\
& & 0 & 1 & 0 & & & & \\
& & -1 & 1 & 0 & & & & \\
& & & -1 & 1 & 0 & & & \\
& & & & 0 & 1 & 0 & & \\
& & & & & 0 & 1 & 0 & \\
& & & & & & 0 & 1 & \\
\end{pmatrix}
Q = w_{eq}(R - I)
\begin{pmatrix}
0 \\
0 \\
0 \\
L^{(1)} + L^{(4)} \\
L^{(2)} + L^{(5)} + L^{(7)} \\
L^{(3)} + L^{(6)} + L^{(8)} \\
0 \\
0 \\
0 \\
\end{pmatrix}
\tag{9.17}
$$

Solving Eq. 9.17 provides the volumetric discharge for each cell of the catchment for the equilibrium condition. It is easily shown that this conserves mass.

To see how this formulation accumulates flow, we take the example of an impervious plane where the equivalent width is 1,000 m.

Example Watershed Parameters

(R–I) = 3.6 mm/h	Rainfall excess
L = 1000 m	Resolution
Sum (L) = 8000 m	8 finite elements drain to this cell
w = A/L = 1000 m	8 cells/8 finite elements

Solving Eq. 9.17 for these constants, we obtain a solution vector containing the flow rates, Q, at each node. For these constants, we obtain

$$
\begin{pmatrix}
Q_1 \\
Q_2 \\
Q_3 \\
Q_4 \\
Q_5 \\
Q_6 \\
Q_7 \\
Q_8 \\
Q_9 \\
\end{pmatrix}
=
\begin{pmatrix}
0 \\
0 \\
0 \\
2 \\
5 \\
8 \\
0 \\
0 \\
0 \\
\end{pmatrix}
\tag{9.18}
$$

Conservation of mass is preserved because there are eight cells draining to the outlet cell (node 6). Because the system is at equilibrium, input equals output and we obtain

$$Q_8 = \frac{3.6}{3600 * 1000} * 8 * 1000^2 = 8 \tag{9.19}$$

In this formulation, upstream cells produce an inflow of 8 m^3s^{-1} rather than 9 m^3s^{-1}. Thus, the assembly of the finite elements has been demonstrated using the residual equation to form the global system and its solution is demonstrated to conserve mass.

9.4   Time-Dependent Solution

To obtain the time-dependent solution, a time-marching scheme is required to solve Eq. 9.10. To solve this system, a discretization scheme is needed, such as the finite difference scheme. It should produce accurate and stable solutions. The finite difference formulation is

$$CA_{new} = CA_{old} - \Delta t S[(1 - \theta)Q_{old} + \theta Q_{new}] + \Delta t [(1 - \theta)F_{old} + \theta F_{new}] \tag{9.20}$$

where θ is the weighting coefficient, Δt is the time step, and the *new* and *old* subscripts denote the value of the variable at the current and next time step. The explicit solution ($\theta = 0$) has been chosen because it is faster than implicit schemes even with the limiting time step of the Courant condition. The time step is a function of the Courant condition, which requires that the time step be less than the time of travel for a gravity wave to propagate across the element at celerity equal to $\sqrt{gh}$ at an equilibrium flow rate. The time step associated with the courant condition is given as

$$\Delta t = L/\sqrt{gh} \tag{9.21}$$

where L is the length of the smallest element. The most restrictive element can be identified as that which has the shortest Courant time step. In practice this is difficult because the spatially variable rainfall of any given event may force different elements to be the most restrictive.

9.5   Rainfall Excess Determination

The right-hand forcing vector utilizes the rainfall excess at each time step. Rainfall rate minus infiltration rate is the rainfall excess in each grid. The potential infiltration rate is computed using the Green and Ampt equation introduced in Chap. 5. Comparison of rainfall rate to the potential rate determines whether all the rain infiltrates or whether the excess is available for routing to the next downstream cell. Any excess arriving from upstream at a particular cell location is added to the rainfall for that cell. In this way, run-on from upstream may infiltrate or add to the rainfall excess in each cell. Comparison of rainfall rate to the potential rate determines whether all the rain infiltrates or whether the excess is available for routing to the next downstream cell.

During the storm event and afterward for a specified monitoring period, the Green and Ampt equation computes the amount of water infiltrating. As long as the rainfall rate i is less than the potential infiltration, the cumulative infiltration is simply equal to

$$F(t + \Delta t) = F(t) + i\Delta t \tag{9.22}$$

Then, when ponding occurs, i.e., as soon as it exceeds the potential infiltration, water starts ponding at the surface and then

$$F_P = \frac{K_e \psi_f \theta_d}{i - K_e} \quad (K_e < i) \tag{9.23}$$

Afterward, the infiltration rate is somewhat less than the potential infiltration rate and is defined by

$$f(t) = K_e \left(\frac{\psi_f \theta_d}{F(t)} + 1 \right) \quad \text{if } F(t) \neq 0 \tag{9.24}$$

where the cumulative infiltration $F(t)$ is

$$F(t + \Delta t) = F(t) + K_e \Delta t + \psi_f \theta_d \ln \left(\frac{F(t + \Delta t) + \psi_f \theta_d}{F(t) + \psi_f \theta_d} \right) \tag{9.25}$$

Equation 9.25 can be solved at each time step using a fixed-point method. Assuming that the time discretization is sufficiently small, Newton iteration should converge. We define the functional $G(x)$ as

$$G(x) = x - F(x) = x - \psi_f \theta_d \ln \left(\frac{x + \psi_f \theta_d}{F(t) + \psi_f \theta_d} \right) - K_e \Delta t \tag{9.26}$$

where $x = F(t + \Delta t)$. The form of Eq. 9.26 allows us to write it for an iterative solution:

$$G'(x^j) = \frac{0 - G(x^j)}{x^{j+1} - x^j} \tag{9.27}$$

where x^j is the cumulative infiltration amount identified at the jth iteration of the Newton's formula, at time $t + \Delta t$. Taking the derivative of Eq. 9.26, we obtain

$$G'(x) = \frac{x}{x + \psi_f \theta_d} \tag{9.28}$$

Introducing Eq. 9.28 into Eq. 9.27 provides the final Newton's iteration computed at each time step, i.e., the cumulative infiltration quantity:

Fig. 9.6 Representation of the flow depth computed in each grid cell incorporating the spatially variable terrain, soils, land use/cover, and rainfall input of the Blue River basin, Oklahoma

$$x^{j+1} = -\psi_f\theta_d + \frac{x^j + \psi_f\theta_d}{x^j}\left\{F(t) + K_e\Delta t + \psi_f\theta_d \ln\left(\frac{x^j + \psi_f\theta_d}{F(t) + \psi_f\theta_d}\right)\right\} \quad (9.29)$$

A substitution method can also be used but experience shows Newton's iteration converges faster.

Building a digital watershed from a 3×3 kernel permits the solution of the flow analogy for the entire watershed. An example of a distributed runoff map produced at 270-m resolution for the 1,200 km^2 Blue River basin is shown in Fig. 9.6. The flow depth is draped on a 3D rendering of the digital terrain model. Darker areas represent deeper flow depths and lighter areas more shallow. The drainage network is apparent in this image because of the accumulation of runoff downstream of the more intense rainfall. There are areas where there is no runoff caused by low rainfall rates during this particular time period and/or soils that have infiltration rates that exceed the rainfall rate. The map shown is for a particular time interval of 1 h during an event simulated using radar rainfall input.

9.6 Subsurface Flow

Liuzzo et al. (2009) investigated interactions between surface and subsurface runoff using the TIN-based Real-Time Integrated Basin Simulator model (tRIBS) through continuous simulation of synthetically generated precipitation. Arnold et al. (1993)

and Arnold et al. (2000) discussed a comprehensive approach to modeling surface and groundwater coupling and regional estimation of base flow and recharge. An infiltration model can be used to model the flux arriving at the bottom of the soil profile and the top of a water table or an aquifer. Then the aquifer model is used to estimate discharge to streamflow, which then must utilize routing and diffuse runoff to make the necessary connections to the measured streamflow downstream. Barrett and Charboneau (1997) used this approach to model groundwater discharge from the Edwards Aquifer to Onion Creek and interior sub-watersheds, located in Central Texas, USA. Edwards is a karst aquifer with high porosity and is directly connected to the surface by infiltration and to stream channels where whirlpools form as streamflow enters the aquifer storage directly. One difficulty they found was that rainfall was not adequately measured over the aquifer recharge area.

Groundwater recharge consists of water that infiltrates into the ground to a depth below the root zone reaching the water table and forming an addition to the groundwater reservoir (De Vries and Simmers 2002). The complexity of recharge to an unconfined aquifer is determined primarily by the occurrence, duration, and intensity of precipitation, which penetrates the soil directly and percolates to the saturated zone. Other factors influencing recharge include evapotranspiration from the soil and vegetation, which limits the amount of recharge derived from rainfall and subsequent infiltration. Methods to measure or estimate recharge include physical, chemical, and numerical techniques (Scanlon et al. 2003). Most physical and chemical methods provide detailed information about recharge rates and spatial distribution but are very laborious, highly site-specific, and expensive to apply and monitor. On the other hand, numerical methods provide estimates of recharge rates over large areas but its accuracy depends upon model calibration and temporal and spatial scales of measured data (Scanlon et al. 2003). Valett and Sheibley (2009), Kollet and Maxwell (2006), and Kalbus et al. (2006) described the numerical methods used to quantify the interaction between surface and subsurface water. A fully integrated model, combining groundwater into a surface water model at the watershed scale, was presented by Arnold et al. (1993). Alternatively, Scibek et al. (2007) described the integration of surface water *within* a subsurface model.

Subsurface runoff can be seen in the streamflow response even where significant aquifer or karst features are not present. Shallow water tables within the soil profile or parent material can intercept infiltration and directly discharge to streamflow. This process is referred to as interflow or quick return flow. A thorough description of the possible interactions of surface and groundwater was presented by Sophocleous (2002). The significance of these interactions relates to groundwater exploitation as it corresponds to safe yield and groundwater depletion. Further distinctions can be made depending on the pathway through which infiltration augments a water table thus accelerating the flux to a stream. The subsurface runoff component is particularly important in humid and subhumid climates, karst watersheds, and in mountain watersheds where streamflow is dominated by the delayed runoff traveling in the subsurface through these mechanisms. Sloan and Moore (1984) provided a comprehensive description of the mechanism and modeling of stormflow on a steeply sloping mountainous watershed. Tachikawa and

Takasao (1996) incorporated the subsurface runoff in a distributed rainfall-runoff model that relied on Triangulated Irregular Networks, or TINs, which was later extended by Lee et al. (2009). Their approach dealt with water movement on a slope element that connects downstream to channel elements. The simulation of this flux uses a surface–subsurface kinematic wave equation driven by the change in slope of the water table (Takasao and Shiiba 1988). A fundamental assumption of this and other such approaches is that the top layer of the soil profile has an infiltration rate greater than the rainfall rate, thus insuring that there is no overland runoff and only subsurface runoff can occur.

Because spatially and temporally variable rainfall is a key driver of groundwater recharge, Moreno and Vieux (2013) captured the rainfall variability with radar and rain gauge measurements for model input to a distributed hydrologic model of the Blue River Basin in south central Oklahoma, USA. Figure 9.7 shows three maps of total rainfall from 1994–2007. Rainfall ranges from 1000 mm to over 1700 mm during this period with considerable more variation than shown by the rain gauge map. This variability over a relatively short distance, a few hundred kilometers, emphasizes the importance of spatial variation in rainfall depth used in modeling recharge. Within the basin's upper reaches, directly connected karst features receive recharge and then discharge via springs to the river. With representative high-resolution GARR rainfall inputs over a 12-year period, the streamflow record was extended back to 1994 through modeling of surface and subsurface runoff and groundwater recharge. The streamflow derives from direct surface runoff and aquifer discharge from karst geology of the Arbuckle-Simpson anticline (Fairchild et al. 1990 and Faith et al. 2010). The nearly continuous streamflow and unique character led to its designation as an Oklahoma Scenic River because it is unimpeded by man-made reservoirs and with adequate base flow, it is stocked with trout during fall and winter months because of the abundant supply from groundwater.

A comprehensive water resource assessment of the Arbuckle-Simpson aquifer within the Blue River was presented by Osborn (2009). An overview of the distributed modeling approach to streamflow and aquifer recharge was described in Moreno and Vieux (2013). One of two stream gauges is located in the Blue River near Connerville (USGS 07332390) comprising 420 km^2. At the time of the study, this stream gauge only had a few years of streamflow record. The location of the Connerville gauging station was well located for capture of discharge measurements from springs located at the contact between the karstic Arbuckle group and the Tishomingo Granite. The second, most downstream gauge, Blue River (USGS 07332500), is located about 86 km downstream of Connerville, comprising 1,235 km^2.

The modeling approach computed time series of runoff and grids of infiltration by means of a physics-based distributed hydrologic model, V$flo^{®}$. The 200-m resolution of the model allows accounting for infiltration excess and saturation excess occurring at a small scale. Grids of infiltration produced for the area upstream of Connerville are subsequently combined with grids of actual evapotranspiration from the water balance, to estimate groundwater recharge. Groundwater recharges the *epikarst* and a recharge that reaches the water table is determined by spatially variable rainfall, soils, and land use affecting the partition between runoff and

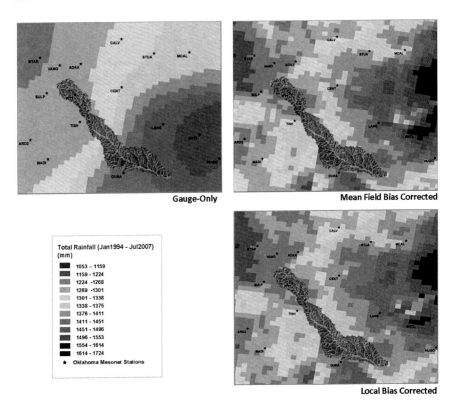

Fig. 9.7 Rainfall total maps (1994–2007) produced using rain gauges (*stars*) interpolated to give the gauge-only rainfall map *upper left* and from a gauge-corrected radar by the Mean Field Bias and Local Bias methods (Moreno and Vieux 2013)

infiltration. Validation of these estimates is accomplished by comparing the modeling results on an annual basis with the automated base flow separation technique, called PART (Rutledge 1998), a program developed by USGS for partitioning direct runoff and base flow. During the period when streamflow was available at Connerville, the difference between PART and V*flo*® estimates of recharge ranged from 8 % in 2004 to 27 % in 2006, with a mean of 15 %. While not in perfect agreement, the result from using two different methods was encouraging and gave confidence in the distributed model approach

9.7   Subsurface and Surface Water Balance

In summary, the results of the Moreno and Vieux (2013) study demonstrates the importance of the temporal variability of rainfall and its control on the amount of recharge in the Arbuckle group in the upper Blue River. Further, it is worth noting the variation of rainfall over the area, both spatially and over the study period.

Table 9.2 Water balance and recharge computed for the Blue River

Component	1994	1995	1996	1997	1998	1999	2000	2001	2002	2003	2004	2005	2006	Avg
Precip. (P) (mm)	1,016	861	1,095	1,019	930	932	1,021	622	1,049	678	1,029	884	820	920
Runoff (mm)	30	25	15	33	109	10	71	147	1	1	56	104	43	50
Infiltration (mm)	947	826	922	970	749	912	930	556	1,024	668	1,013	828	742	853
Actual ET (mm)	683	508	815	818	660	754	876	384	693	589	762	561	622	671
Recharge (mm)	264	318	107	152	89	157	53	173	330	79	251	267	119	182
Gw/P	0.3	0.4	0.1	0.1	0.1	0.2	0.0	0.3	0.3	0.1	0.2	0.3	0.1	0.2

In Table 9.2, the water balance components are presented for 1994–2006 and as annual averages (Avg). Rainfall totals varied from as low as 622 mm to over 1,095 mm over the Blue River watershed and recharge as a fraction of precipitation (Gw/P), ranged from 0.0 to 0.4.

9.8   Summary

Physics-based distributed modeling achieves a close coupling of runoff generation and the routing of this runoff through a drainage network. This approach is achieved through a *hydraulic* approach to *hydrology*, which is a departure from traditional methods where runoff generation and routing are artificially separated. The mathematical analogy and numerical algorithms described are the foundation for modeling runoff at any location in a watershed described with a drainage network.

From a DEM, the derivative flow direction and slope are used to solve the kinematic wave equations. The direction of principal slope classified according to the eight neighboring cells defines the connectivity used to assemble a system of equations. Using the finite element method to solve the kinematic wave analogy and GIS maps of parameters with rainfall input, the runoff process is simulated from hillslope to the river basin. The next step is to produce runoff hydrographs and output maps of flow depth and cumulative infiltration that agree with observed hydrologic variables.

Distributed modeling of both surface and subsurface runoff depends on capturing the variation of rainfall, topography, soils, and land use/cover and modeling the flux through a drainage network at the surface and through subsurface representation of an aquifer or shallow water table. The sensitivity of runoff and recharge driven by rainfall demonstrates the need for accurate representation of inputs and model parameters, which in turn determine the accuracy of the surface runoff and recharge contributing to streamflow. Given specific climatic and geologic settings, recharge may be controlled by rainfall and/or evapotranspiration. Modeling of recharge and streamflow was demonstrated for a karst watershed using a distributed physics-based approach that allowed the extension of streamflow records and thereby developed a more reliable water balance than could be achieved solely from measurements of streamflow and precipitation alone. In the next chapter, we examine the calibration of parameter maps used to parameterize a distributed model.

References

Abbott, M.B., J.C. Bathurst, J.A. Cunge, P.E. O'Connell, and J. Rasmussen. 1986a. An introduction to European hydrological system—Systeme Hydrologique Europeen, SHE, 1 History and philosophy of a physically-based distributed modeling system. *Journal of Hydrology* 87: 45–59.

Abbott, M.B., J.C. Bathurst, J.A. Cunge, P.E. O'Connell, and J. Rasmussen. 1986b. An Introduction to the European Hydrological System—Systeme Hydrologique European, SHE, 2: Structure of a physically-based distributed modelling system. *Journal of Hydrology* 87: 61–77.

Arnold, J.G., P.M. Allen, and G. Bernhardt. 1993. A comprehensive surface-groundwater flow model. *Journal of Hydrology* 142: 47–69.

Arnold, J.G., R.S. Muttiah, R. Srinivasan, and P.M. Allen. 2000. Regional estimation of baseflow and groundwater recharge in the upper Mississippi River Basin. *Journal of Hydrology* 227: 21–40.

Barrett, M.E., and Charbeneau, R.J., 1997. A parsimonious model for simulating flow in a karst aquifer. *Journal of Hydrology* 196(1): 47–65.

Chow, V.T., D.R. Maidment, and L.W. Mays. 1988. *Applied hydrology*. New York: McGraw-Hill.

Clark, M.P., B. Nijssen, J.D. Lundquist, D. Kavetski, D.E. Rupp, R.A. Woods, J.E. Freer, E.D. Gutmann, A.W. Wood, L.D. Brekke, and J.R. Arnold. 2015a. A unified approach for process-based hydrologic modeling: 1. *Modeling concept. Water Resources Research* 51(4): 2498–2514.

Clark, M.P., B. Nijssen, J.D. Lundquist, D. Kavetski, D.E. Rupp, R.A. Woods, J.E. Freer, E.D. Gutmann, A.W. Wood, D.J. Gochis, and R.M. Rasmussen. 2015b. A unified approach for process-based hydrologic modeling: 2. Model implementation and case studies. *Water Resources Research* 51(4): 2515–2542.

De Vries, J.J., and Simmers, I., 2002. Groundwater recharge: an overview of processes and challenges. *Hydrogeology Journal* 10(1): 5–17.

Downer, C.W., and F.L. Ogden. 2004. GSSHA: a model for simulating diverse streamflow generating processes. *Journal of Hydrologic Engineering* 9(3): 161–174.

Fairchild, R. W., Hanson R. L., and Davis R. E., 1990. *Hydrology of the Arbuckle Mountains Area, South-Central Oklahoma*, Circular 91, Oklahoma Geological Survey-OGS, 112 pages.

Faith, J.R., Blome, C.D., Pantea, M.P., Puckette, J.O., Halihan, T., Osborn, N., Christenson, S., and Pack, S., 2010. Three-dimensional geologic model of the Arbuckle-Simpson Aquifer, south-central Oklahoma (No. 2010–1123). US Geological Survey.

Gochis, D.J., W. Yu, and D.N. Yates. 2014. The WRF-Hydro model technical description and user's guide, version 2.0. NCAR Technical Document.

Ivanov, V.Y., E.R. Vivoni, R.L. Bras, and D. Entekhabi. 2004a. Catchment hydrologic response with a fully distributed triangulated irregular network model. *Water Resources Research*, 40(11).

Ivanov, V.Y., E.R. Vivoni, R.L. Bras, and D. Entekhabi. 2004b. Preserving high-resolution surface and rainfall data in operational-scale basin hydrology: a fully-distributed physically-based approach. *Journal of Hydrology* 298(1): 80–111.

Jayawardena, A.W., and J.K. White. 1977. A finite element distributed catchment model (1): Analysis basis. *Journal of Hydrology* 34: 269–286.

Jayawardena, A.W., and J.K. White. 1979. A finite element distributed catchment model (2): Application to real catchment. *Journal of Hydrology* 42: 231–249.

Jain, M.K., and Vijay P. Singh. 2005. DEM-based modelling of surface runoff using diffusion wave equation, *Journal of Hydrology* 302(1–4): 107–126, 1 February 2005, http://dx.doi.org/10.1016/j.jhydrol.2004.06.042.

Julien, P.Y., and B. Saghafian. 1991. CASC2D user manual—a two dimensional watershed rainfall-runoff model. Civil Engineering Rep. CER90-91PYJ-BS-12, Colorado State University, Fort Collins: 66.

Kalbus, E., F. Reinstorf, and M. Schrimer. 2006. Measuring methods for ground water–surface water interactions: A review. *Hydrology and Earth Systems Sciences* 10: 873–887.

Kazezyılmaz-Alhan, C., and M. Medina Jr. 2007. Kinematic and diffusion waves: Analytical and numerical solutions to overland and channel flow. *Journal of Hydraulic Engineering* 133:2 (217), 217–228, 10.1061/(ASCE)0733-9429.

Kollet, S.J., and R.M. Maxwell. 2006. Integrated surface-groundwater flow modeling: a free-surface overland flow boundary condition in a parallel groundwater flow model. *Advanced Water Resources* 29: 945–958.

Koren, V., S. Reed, M. Smith, Z. Zhang, and D.-J. Seo. 2004. Hydrology laboratory research modeling system (HL-RMS) of the US national weather service. *Journal of Hydrology* 291(3–4): 297–318.

Kuchment, L.S., V.N. Demidov, and I.G. Motovilov. 1983. *Formirovanie rechnogo stoka: fiziko-matematicheskie modeli*. Nauka.

Kuchment, L.S., V.N. Demidov, and Y.G. Motovilov. 1986. A physically based model of the formation of snowmelt and rainfall runoff. In: Modelling Snowmelt-Induced Processes (Proceedings of Budapest Symposium, July 1986), IAHS Publ., 155:27–36.

Lee, G., Y. Tachikawa, and K. Takara. 2009. Interaction between Topographic and Process Parameters due to the Spatial Resolution of DEMs in Distributed Rainfall-Runoff Modeling. *Journal of Hydraulic Engineering*, 10.1061/(ASCE)HE.1943-5584.0000098, 1059–1069.

Lighthill, F.R.S., and G.B. Whitham. 1950. On Kinematic Waves, I, Flood Measurements in Long Rivers. *Proceedings of the Royal Society of London* 229: 281–316.

Liuzzo, L., Noto, L.V., Vivoni, E.R., and La Loggia, G., 2009. Basin-scale water resources assessment in Oklahoma under synthetic climate change scenarios using a fully distributed hydrologic model. *Journal of hydrologic engineering* 15(2):107–122.

Meadows, M.E., and G.E. Blandford. 1990. Finite element simulation of nonlinear kinematic surface runoff. *Journal of Hydrology JHYDA 7,119*(1/4).

Moreno, M.A., and B.E. Vieux. 2013. Estimation of spatio-temporally variable groundwater recharge using a rainfall-runoff model. *Journal of Hydrologic Engineering* 18(2): 237–249.

Ogden, F.L., and P.Y. Julien. 1994. Runoff model sensitivity to radar rainfall resolution. *Journal of Hydrology* 158: 1–18.

Osborn, N.I. 2009. *Arbuckle-Simpson Hydrology Study: Final Report to the US Bureau of Reclamation*. Oklahoma Water Resources Board.

Ross, B.B., D.N. Contractor, and V.O. Shanholtz. 1979. A finite element model of overland and channel flow for accessing the hydrologic impact of land use change. *Journal of Hydrology* 41: 11–30.

Rutledge, A.T. 1998. Computer Programs for Describing the Recession of Ground-Water Discharge and for Estimating Mean Ground-Water Recharge and Discharge from Streamflow Records-Update. U.S Geological Survey, Water Resources Investigations Report 98-4148, 43 p.

Scanlon, B.R., A. Dutton, and M.A. Sophocleous. 2003. *Groundwater recharge in Texas*. Board, Austin: Texas Water Dev.

Scibek, J., D.M. Allen, A.J. Cannon, and P.H. Whitfield. 2007. Groundwater-surface water interaction under scenarios of climate change using a high-resolution transient groundwater model. *Journal of Hydrology* 333: 165–181.

Segerlind, L.J. 1984. *Applied Finite Element Analysis*, 2nd ed. New York: Wiley.

Singh, V.P. and Woolhiser, D.A., 2002. Mathematical modeling of watershed hydrology. *Journal of hydrologic engineering* 7(4): 270–292.

Sloan, P.G., and I.D. Moore. 1984. Modeling subsurface stormflow on steeply sloping forested watersheds. *Water Resources Research* 20(12): 1815–1822. December.

Sophocleous, M. 2002. Interactions between groundwater and surface water: the state of the science. *Hydrogeology Journal* 10: 52–67.

Tachikawa, Y., and T. Takasao. 1996. HydroGIS 96: Application of Geographic Information Systems in Hydrology and Water Resources Management (Proceedings of the Vienna Conference, April 1996). IAHS Publ. no. 235.

Takasao, T., and M. Shiiba. 1988. Incorporation of the effect of concentration of flow into the kinematic wave equations and its applications to runoff system lumping. *Journal of Hydrology* 102: 301–322.

Tromble, E., R. Kolar, K. Dresback, Y. Hong, B. Vieux, R. Luettich, J. Gourley, K. Kelleher, and Van Cooten, S. 2010. Aspects of Coupled Hydrologic-Hydrodynamic Modeling for Coastal Flood Inundation. Estuarine and Coastal Modeling. Estuarine and Coastal Modeling (2009), ed. Malcolm L. Spaulding, 11th International Conference on Estuarine and Coastal Modeling, Seattle, Washington, November 4–6, 724–743. doi:10.1061/41121(388)42.

Valett, H.M., and R.W. Sheibley. 2009. Ground water and surface water interaction. In *Encyclopedia of Inland Waters*, ed. G.E. Likens, 691–702. Oxford: Academic Press.

V*flo*® User Manual, 2016. Available at: http://vflo.vieuxinc.com/vflo-guide.

Vieux, B.E. 1988. Finite Element Analysis of Hydrologic Response Areas Using Geographic Information Systems. Department of Agricultural Engineering, Michigan State University. A dissertation submitted in partial fulfillment for the degree of Doctor of Philosophy.

Vieux, B.E., V.F. Bralts, L.J. Segerlind, and R.B. Wallace. 1990. Finite element watershed modeling: One-dimensional elements. *Journal of Water Resources Planning and Management*, 116(6): 803–819.

Vieux, B.E., and Segerlind, L.J. 1989. Finite element solution accuracy of an infiltrating channel. In: Finite element analysis in fluids, Proceedings of the Seventh International Conference on Finite Element Methods in Flow Problems, April 3–7, 1989, ed. Chung, T.J., and Karr, G.R., University of Alabama in Huntsville Press. ISBN 978-0942166019, pp: 1337-1342.

Vieux, B.E., and N. Gauer. 1994. Finite element modeling of storm water runoff using GRASS GIS. *Microcomputers in Civil Engineering* 9(4): 263–270.

Vieux, B.E. 2001. *Distributed Hydrologic Modeling Using GIS, ISBN 0-7923-7002-3*, 1st ed. Norwell, Massachusetts, Water Science Technology Series: Kluwer Academic Publishers. 38.

Vieux, B.E., J.E. Vieux. 2002. V*flo*®: A real-time distributed hydrologic model. In Proceedings of the 2nd Federal Interagency Hydrologic Modeling Conference, July 28–August 1, 2002, Las Vegas, Nevada. Abstract and paper on CD-ROM.

Vieux, B.E., C. Chen, J.E. Vieux, and K.W. Howard. 2003. Operational deployment of a physics-based distributed rainfall-runoff model for flood. In Weather Radar Information and Distributed Hydrological Modelling: Proceedings of an International Symposium (Symposium HS03) Held During IUGG 2003, the XXIII General Assembly of the International Union of Geodesy and Geophysics: at Sapporo, Japan, from 30 June to 11 July, 2003. No. 282. International Assn of Hydrological Sciences, 2003.

Vivoni, E.R., V.Y. Ivanov, R.L. Bras, and D. Entekhabi. 2004. Generation of triangulated irregular networks based on hydrological similarity. *Journal of Hydrologic Engineering* 9(4): 288–302.

Vivoni, E.R., D. Entekhabi, R.L. Bras, and V.Y. Ivanov. 2007. Controls on runoff generation and scale-dependence in a distributed hydrologic model. *Hydrology and Earth System Sciences Discussions* 11(5): 1683–1701.

Wigmosta, M.S., L.W. Vail, and D.P. Lettenmaier. 1994. A distributed hydrology-vegetation model for complex terrain. *Water Resources Research* 30(6): 1665–1679.

Chapter 10
Distributed Model Calibration

Ordered Physics-Based Adjustment

Abstract We show that given certain constraints, such as the spatial pattern of parameters, unique solutions do exist, thus making it possible to calibrate a distributed model. As we have shown in previous chapters, the drainage length, slope, and other parameters extracted from DEMs and geospatial data are *resolution dependent*. Thus, it is more likely than not, such parameters would require some adjustment. This chapter presents a method for calibrating a distributed model consistent with the conservation equations that underlie PBD models.

10.1 Introduction

Distributed models are parameterized by deriving estimates of parameters from physical properties, viz., databases of soil properties are used to derive infiltration parameters. Without calibration, a distributed hydrologic model may be used with a priori values that are considered typical of the watershed land surface, vegetation, and soils. Empirically based models cannot be used without calibration because parameter values must be estimated from historic streamflow data. A comprehensive treatment of rainfall-runoff model calibration may be found in Duan et al. (2003) and Wagener et al. (2004). With any model, its success depends on the model structure, how well parameters can be identified using observations to constrain the parameter search algorithm and regionalization of the parameters achieving reduced dimensionality of the optimization problem. Conceptual rainfall-runoff (CRR) models generally have a large number of parameters, which are not directly measurable and must therefore be estimated through model calibration, i.e., by fitting the simulated outputs of the model to the observed outputs of the watershed by adjusting parameters. A great deal of effort has been expended on automatic calibration of CRR models (Gupta et al. 2005) even though the hydrographs generated and associated parameter estimates may not result in acceptable value ranges (Cheng et al. 2002). Agreement between the simulated and observed outputs is measured in terms of a calibration criterion or objective function. The goal of calibration is to find those values for the model parameters that minimize differences according to the objective function.

© Springer Science+Business Media Dordrecht 2016

B.E. Vieux, *Distributed Hydrologic Modeling Using GIS*,
Water Science and Technology Library 74, DOI 10.1007/978-94-024-0930-7_10

The more commonly used automatic calibration techniques rely on direct-search optimization algorithms, such as the simplex method of Nelder and Mead (1965) and the pattern search method of Hooke and Jeeves (1961). These algorithms are designed to solve single-optimum problems and are not able to deal effectively with problems like region of attraction, minor local optima, roughness, sensitivity, and shape. The shuffled complex evolution method developed at the University of Arizona (SCE-UA) is based on a synthesis of the best features from several existing methods plus complex shuffling (Duan et al. 1992, 1993). Designed specifically for the problems encountered in conceptual watershed model calibration, it consists in four concepts: (1) combination of deterministic and probabilistic approaches; (2) systematic evaluation of a 'complex' of points spanning the parameter space, in the direction of global improvement; (3) competitive evolution; and (4) complex shuffling. For this method to perform optimally, these parameters must be chosen carefully. Duan et al. (1994) recommended values for algorithmic parameters based on the results of several experimental studies in which the SAC-SMA model for river and flood forecasting was calibrated by the NWS using different algorithmic parameter setups. CRR models require *tuning* of parameters across a wide range of values because values of the parameter are unknown and not related to physical quantities.

Long-standing development of automated computer-based calibration methods has focused mainly on (1) definition of single or multi-parameter objective functions and (2) automatic optimization algorithms (Sorooshian and Dracup 1980). Yapo et al. (1996, 1998) extended the single-objective function method to a multi-objective complex evolution (MOCOM-UA) capable of exploiting the observed time series. This method is an extension of the SCE-UA single-objective global optimization algorithm. Regardless of the search algorithms used to calibrate CRR models, parameter interaction and parameter stability between storms and inter-annually are still problematic. Boyle et al. (2001) compared simulation of the Blue River basin with lumped and semi-distributed representations and found no improvement using spatially distributed parameters. This could be attributed to the CRR model structure that demonstrates limited prediction accuracy improvement. Lumping of parameter values reduces prediction accuracy and can introduce systematic bias.

While fully distributed physics-based modeling avoids issues with lumping, it should be noted that distributed models can suffer from other sensitivities, more so than lumped models. Vieux et al. (2004) found that hail contamination and bias in NWS QPE radar estimates provided as part of the DMIP experiments (Smith et al. 2004, 2012) could cause significant errors in hydrologic response, while the lumped SAC-SMA models showed little sensitivity. When Looper et al. 2012 subjected the operational MPE to further quality control and bias correction using rain gauges not used in operation, significant improvements in model prediction accuracy are resulted. Smith et al. (2004) reported extensive performance statistics for a range of distributed and lumped CRR models that relied on operational MPE. Lumped models such as the SAC-SMA were less sensitive to input errors as demonstrated by CRR performance with operational MPE as input. One interpretation is to consider that a watershed model functions as a low-pass filter (Oudin et al. 2004) where high-frequency perturbations in model input, potential evapotranspiration (PE), or precipitation are smoothed out. The models tested showed less sensitivity to PE than precipitation, which could be attributed to the way in which TOPMODEL produces runoff, largely due to saturation excess. On the contrary, Vieux et al. (2009) found greater sensitivity to soil moisture than precipitation bias. In any case, for the two watersheds tested, the initial saturation of the soil affects prediction accuracy more than the uncertainty caused by model parameters or gauge-only input.

10.2   Calibration Approach

Physics-based distributed model calibration differs from CRR calibration in two important ways. First, some scheme must be devised to adjust the grid-cell parameters affecting the output. Second, because of the governing equations derived from the physics of conservation of mass and momentum, the parameters should exhibit expected behavior. Lumped models typically suffer from unexpected parameter interaction. Tuning one parameter affects another in ways not anticipated, making the recovery of unique optimal parameter sets doubtful.

Several aspects of the calibration process are of particular importance to distributed models: (1) maps of parameters derived from geographic information system data or remote sensing (GIS/RS) provide spatial distribution; (2) parameters may be scale-dependent because of sampling characteristics of the GIS/RS source; (3) slope and drainage length are dependent on DEM resolution; (4) continuous calibration is useful in reducing uncertainty in soil moisture and its effect on runoff events; and (5) calibration is used to adjust initial parameter estimates from soil properties, DEM, and land use/cover. The agreement between the observed and simulated volume and peak flow may be expressed in terms of a bias or departure. The bias indicates systematic over or under prediction. The departure, whether expressed as an average difference, percentage error, coefficient of determination, or as a root-mean-square error, serves as a measure of the prediction accuracy.

A PBD model has the advantage of having expected parameter response and interaction. Whether manual or automatic methods are employed, a physics-based model capitalizes on the underlying conservation equations. Manual adjustment also profits from the physical relationship, physical significance, and expected response to adjustment of parameters derived from physical properties. To summarize the approach,

1. Estimate the spatially distributed parameters from physical properties.
2. Assign channel hydraulic properties based on measured cross sections where available.
3. Study model sensitivity for the particular watershed.

   3.1. Identify response sensitivity to each parameter.

3.2. Run the model for a continuous period or at least for a range of storms from small, medium, to large events.

3.3. Observe the characteristics of the hydrograph over the range of storm sizes.

3.4. Observe any consistent volume bias.

4. Identify seasonal effects that may influence radar estimation of rainfall, land use/cover, or other factors.

5. Identify any systematic bias due to radar, soil moisture, or hydraulic conductivity.

6. Derive a range of responses for a given change in a parameter, e.g., soil moisture.

7. Categorize and rank parameter sensitivity according to response magnitude.

8. The optimum parameter minimizes the respective objective function.

9. Match volume by adjusting hydraulic conductivity.

10. Match peak by adjusting overland flow roughness.

11. Match time to peak by adjusting channel roughness.

12. Readjust hydraulic conductivity and hydraulic roughness if necessary.

This sequence is based on the *ordered physics-based parameter adjustment* (OPPA) method after Vieux and Moreda (2003). The expected response and the parameter affecting the model response derive from hydrodynamics embodied in the PBD model. The OPPA procedure outlined above can be stated as *increasing the volume of the hydrograph* is achieved by *decreasing hydraulic conductivity* and similarly, *increasing peak flow* is achieved by *decreasing hydraulic roughness*. The OPPA method is a significant departure from traditional calibration approaches used with CRR hydrologic models. Where CRR parameters have limited physical basis, they may be adjusted over a wide range without regard to physically realistic ranges (cf. chapters in Duan et al. 2003). In contrast, PBD model parameters have a physical basis and remain within physically realistic ranges while adjusted.

Channel parameterization should be applied according to measured cross sections and measured or visual estimates of hydraulic roughness. If this is unavailable, then channel hydraulic characteristics may be estimated from similar channels or local knowledge and then adjusted. Multiple gauging stations within a river basin help resolve timing problems associated with channel hydraulics and aid in the adjustment process. Consistent bias in timing may be related to the channel or the overland flow hydraulics, or both. In either case, these parameters are estimated, and then adjusted to minimize the objective function. Note that if unreasonable channel slopes result from constrained drainage network extraction, then hydraulic roughness or other factors controlling the hydrograph response will be affected (cf. Chap. 7).

Calibration methods designed for CRR models, such as the SCE-UA, do not have governing differential equations to guide the search for optimal solutions. PBD models on the other hand do have governing differential equations that can be used to guide the optimization scheme. Either a grid-based search can be used, or one that employs a gradient to direct the search in parameter space to find the optimum set. Inverse methods leverage the governing differential equations to direct the parameter search, whereas other methods such as the genetic algorithm explore parameter sets by following a lineage of successful parameter sets. Inversion of the kinematic wave analogy (KWA) equations has been developed to identify optimal parameters described in following sections.

10.3   Distributed Model Calibration

Physics-based distributed model calibration differs from lumped calibration in two important ways. First, some scheme is needed to adjust the grid-cell parameters that affect the output. Second, the parameters controlling mass and momentum conservation equations exhibit expected behavior. Calibration of empirical models typically suffers from unexpected parameter interaction where tuning one parameter affects another in unanticipated ways. Empirical models such as the Sacramento Soil Moisture Accounting (SAC-SMA) rely on parameters that have only limited association with available geospatial mapping such as generalized soil maps and thus have difficulties in defining the a priori spatial variation of parameters (see Koren et al. 2000). Without a strong a priori spatial map of parameter variability, calibration efficiency measures may yield poor results (Pokhrel 2007). Furthermore, at downstream locations damping and dispersion of the effects of spatial variability can confound the assignment of model parameter values based solely on hydrograph shape (Pokhrel and Gupta 2010). Pokhrel et al. (2008) suggested that in the presence of model structural and data errors, such as those present in the SAC-SMA model, attempts to calibrate spatially distributed parameter fields can be prone to failure. It should be noted that their approach relied on spatially distributed model versions of the SAC-SMA model rather than a physics-based modeling approach.

This makes it difficult to recover unique optimal parameter values for empirical modeling. Several aspects of the distributed model calibration process are of particular importance:

1. Distributed hydrologic models require calibration of many parameters at many locations but without or with only limited observations.
2. Distributed maps of parameters should be adjusted such that the relative contributions are preserved while the absolute values change.
3. Parameters may require adjustment because they are resolution dependent, e.g., hydraulic roughness.
4. Given uncertainties in soil properties mapped over the river basin, estimated values may only be within an order of magnitude, e.g., saturated hydraulic conductivity.
5. Absolute values may not be known but relative magnitudes can be estimated for specific land uses/cover or soil types.
6. Hydrologic observations are usually limited to stream gauges at the outlet of a river basin or rain gauges scattered across large areas.

These aspects guide the parameter adjustment and calibration of a PBD model.

10.3.1 Parameter Adjustment

Adjusting the distributed parameters is done in a way that minimizes the observed and simulated differences. Because we derive parameters from GIS maps related to soils and infiltration, it is natural to adjust the parameter map values for purposes of calibration. Once the spatial pattern has been established by land use/cover or soil maps, the relative contribution of the parameter is established. Infiltration parameters derived from soils maps represent relative differences in saturated hydraulic conductivity but may not be correct in magnitude. Once we have the spatial pattern established, an adjustment method is needed that preserves this pattern. Adjusting the saturated hydraulic conductivity requires an objective function related to volume to measure progress toward calibrating the model. Similarly, adjustment of hydraulic roughness is guided by an objective function related to peak and timing of the hydrograph.

The parameter adjustment can be accomplished by a multiplicative or additive constant. Scaling of parameter maps by multiplying the values by a factor can cause reduced variance. The mean value can be adjusted up or down but the variance will be exaggerated or diminished. This can easily be demonstrated by taking an array of parameter values and applying first an additive constant and then a multiplicative constant and observing the effect on the variance and mean values. Pokhrel et al. (2008) expanded the multiplier and additive constants to include a power term that introduces nonlinearity. In essence, they confirmed that the OPPA approach by Vieux and Moreda (2003) and Vieux (2004) is a useful method for adjusting the mean of the parameter distribution, even though some distortion of the variance can occur when a large adjustment is applied. Francés et al. (2007) showed that temporal and spatial scale effects appear because the averaging of nonlinear processes introduces scale effects that result in a difference between point process values and average values over a subbasin or even for a grid cell. They observed that due to these effects, the model may be calibrated but may not validate properly for different runoff events. They proposed a means of estimating the multiplicative adjustment necessary to represent scale-dependent values. Of particular importance to calibration of gridded distributed parameter models is that effective parameter values at the macroscale can be less than the mean point parameter value simply due to averaging. Recall the scale-dependency of parameters found by Niedda (2004), which suggests that calibration is specific to grid-cell size. Mendoza et al. (2015) argued that complex process models can be overly constrained by fixed parameters rather than allowing them to vary according to a probabilistic distribution rather than fixing a mean parameter value for each cell. Relaxing a priori parameter constraints helped improve model calibration and simulation.

Given a map of parameter values, adjustment will likely be necessary due to uncertainty in assigned parameter values to a given land use map or infiltration parameters associated with a soil map and due to scale affects associated with assigning point values to grid cells of an arbitrary size. While adjustment of the parameter map can be accomplished by multiplication, adding or subtracting a constant tends to preserve the variance of the parameter map. To avoid changing the parameter variance caused by a multiplicative factor, an additive factor that is equivalent may be derived. An additive factor can be developed as follows. Multiplying the map of K values by α results in a new map, K',

$$K' = \alpha * K \tag{10.1}$$

The expected value of the resulting 'calibrated' map, K', is related to the expected value of the original map, K, by

$$K'_{avg} = \alpha * K_{avg} \tag{10.2}$$

We can now introduce an additive factor, τ, such that K'_{avg} differs from K_{avg} by this factor:

$$\tau = K'_{avg} - K_{avg} \tag{10.3}$$

Note that the mean values are also related by α similar to Eq. (10.1):

$$K'_{avg} = \alpha * K_{avg} \tag{10.4}$$

Substituting Eq. (10.4) into Eq. (10.3), we obtain

$$\tau = K_{avg} * (\alpha - 1) \tag{10.5}$$

In this way, we can adjust the parameter maps by an additive constant that is related to the multiplicative constant by Eq. (10.5). The equation development is shown for α and hydraulic conductivity, K but may be equivalently used for other parameters in the model formulation.

For example, we might need to reduce saturated hydraulic conductivity, K, by an order of magnitude. Using a multiplicative factor of $\alpha = 0.1$ and a mean of 3.31 cm/h, the additive constant given by Eq. (10.5) is

$$\tau = 3.31 * (0.1 - 1) = -2.97$$

As presumed, adjusting K by the multiplicative or additive constant reduces the hydraulic conductive by an order of magnitude and as expected, the multiplicative constant reduces the variance from 7.85 to only 0.08 due to application of the multiplier, 0.1.

The hydraulic conductivity map shown in Fig. 10.1 is derived from soil survey maps and soil properties for a portion of North Carolina encompassing the Tar River basin. The grid is sampled at 1-km resolution and is used in real-time distributed hydrologic modeling. Table 10.1 summarizes the statistics of this hydraulic conductivity map without adjustment ($K*1.0$); with adjustment by a scalar multiplier ($K*0.1$); and adjustment by an additive constant ($K-2.97$). Note that the mean

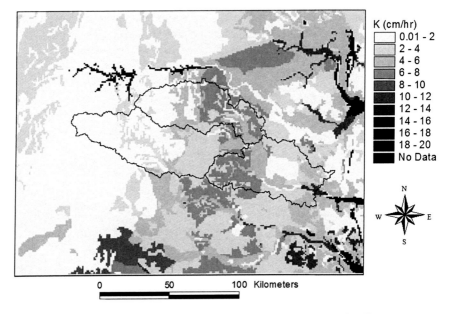

Fig. 10.1 Hydraulic conductivity parameter map for the Tar River, North Carolina

(3.31) is decreased by an order of magnitude as intended by both the multiplicative and additive constant to a value of 0.33 and 0.34, respectively. Instead of reducing, the additive constant preserves the variance, remaining at 7.85 even after adjustment.

For demonstrative purposes, the parameter was allowed to become negative through the additive constant of −2.97. In practice, this parameter must be limited to nonnegative values.

Table 10.1 Summary statistics of hydraulic conductivity without adjustment ($K*1.0$); with adjustment by a scalar multiplier ($K*0.1$); and adjustment by an additive constant ($K–2.97$)

Statistic	Unadjusted ($K*1.0$)	Multiplicative ($K*0.1$)	Additive ($K–2.97$)
Sum	152120	15212	15500
Count	46,000	46,000	46,000
Mean	3.31	0.33	0.34
Maximum	20.13	2.01	17.16
Minimum	0.00	0.00	−2.97
Range	20.13	2.01	20.13
Variance	7.85	0.08	7.85
Standard deviation	2.80	0.28	2.80

10.3.2 Cost Functions

The primary goal of calibration is minimization of model output error by adjusting parameters. The Nash–Sutcliffe statistic is a measure of how closely the simulated time series agrees with observed ones (Nash–Sutcliffe, 1970). The Nash–Sutcliffe term, F_o^2, is the initial variance of the observed record defined as

$$F_o^2 = \Sigma \left(V_{obs} - V_{avg} \right)^2 \tag{10.6}$$

where V_{obs} is an observed quantity such as volume of an event and V_{avg} is the mean of the events. The sum of the square of the differences between observed and simulated volumes, F^2, is defined as

$$F^2 = \Sigma \left(V_{obs} - V_{sim} \right)^2 \tag{10.7}$$

where V_{sim} is the simulated hydrograph integrated over time. The Nash–Sutcliffe statistic, R^2, is

$$R^2 = \left(F_o^2 - F^2 \right) / F_o^2 \tag{10.8}$$

The modified Nash–Sutcliffe is usually expressed as $1\text{-}R^2$. A useful form of the Nash–Sutcliffe Efficiency (NSE) is composed of the correlation, bias, and relative variability between simulated and observed values, which is defined as

$$NSE = 2 \cdot \alpha \cdot r - \alpha^2 - \beta^2 \tag{10.9}$$

where α is $\sigma_{sim}/\sigma_{obs}$; β is $(\mu_{sim} - \mu_{obs})/\sigma_{obs}$; r is the Pearson correlation coefficient; μ and σ are the mean and standard deviation of the observed (obs) and simulated (sim) and observed (obs) series; and the last term, β_n, is the bias between simulated and observed mean values of the time series normalized by the standard deviation of the observed time series. Application of this form of NSE was successful when two inputs to a physics-based distributed model, $Vflo^{®}$, were tested by Looper et al. (2012). Models that have been calibrated using one input (radar) can show bias when those results are compared using a different input (rain gauge). Using just the first two terms of Eq. (10.9), relating to variance, provides a basis for comparing results that contain bias.

Error minimization methods that search the parameter space require a cost function that can be differentiated with respect to each parameter. A general form of the cost function for volume difference is

$$J^2 = \frac{1}{2} \sum \| V_{obs} - V_{sim} \|^2 \tag{10.10}$$

whereas the same cost function can be expressed for discharge as

$$J^2 = \frac{1}{2} \sum \|Q_{obs} - Q_{sim}\|^2 \qquad (10.11)$$

where the summation is over the assimilation period. Before moving to the parameter search algorithms in the next section, it is helpful to observe the behavior of the model in response to adjustment of the input parameters. Hydraulic conductivity affects the volume of runoff and resulting discharge, whereas hydraulic roughness affects the timing and peak discharge. Two parameters that can affect runoff volume in a watershed are soil depth, S_d, saturation excess and saturated hydraulic conductivity, K_{sat}, affecting infiltration rate excess. Looper et al. (2012) calibrated a PBD model by minimizing the difference between observed and simulated streamflow, using a function based on Nash–Sutcliff, E_{Nash} and J_v computed as

$$E_{Nash} = 1 - \frac{\sum_{i-1}^{n} \left(Q_{sim}^i - Q_{obs}^i\right)^2}{\sum_{i-1}^{n} \left(Q_{sim}^i - \bar{Q}\right)} \qquad (10.12)$$

where Q_{sim} is simulated hourly streamflow, Q_{obs} is the observed hourly streamflow, and $\bar{Q}$ is the mean hourly observed streamflow at the ith timestep. Reducing the dimensionality of the problem is accomplished by adjusting the soil depth and saturated hydraulic conductivity by means of multiplicative calibration factors. Using a direct grid search, the two objective functions were tested using a model of the 1,232 km^2 Blue River basin for a continuous record of MPE inputs during 1996–2002. The resulting surfaces are similar with coincident global minima as shown in Fig. 10.2 for E_{Nash} and Fig. 10.3 for J_v.

In this calibration, the a priori Green and Ampt infiltration parameter set was derived from soil maps averaging across the basin with $S_d = 1080$ mm and $K_{sat} = 6.6$ mm/h, whereas both objective functions have global maxima at $S_d = 430$ mm and $K_{sat} = 9.4$ mm/h. The E_{Nash} function is somewhat flatter than J_v but has similar optimal parameter values of K_{sat} ranging from 8.7 to 11.4 mm/h. In either cost surface, the elongated contours along the K_{sat} dimension compared with the S_d dimension suggest more sensitivity to soil depth (saturation excess) than hydraulic conductivity controlling infiltration rate excess (Looper et al. 2012).

As described by the OPPA method, the order of parameter adjustment is important. There is no point in adjusting the timing and peak *before* the volume of runoff is brought into agreement with observed values. Once the runoff volume matches to within some specified range, we can adjust hydraulic roughness to match timing and peak discharge. The volume should be adjusted first, and then the hydraulic roughness to match timing and peak. Channel roughness can then be adjusted to improve the timing of the peak. By adjusting scalar multipliers, $\alpha_0 K$, the volume is adjusted, followed by peak discharge/stage and timing found by adjusting

Fig. 10.2 Cost surface for event volume error based on Nash–Sutcliffe efficiency, E_{Nash}

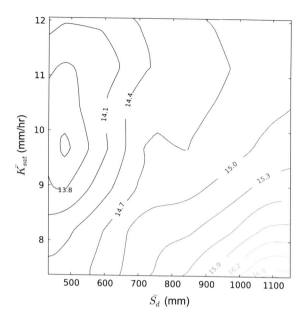

Fig. 10.3 Cost surface for event volume error based on Jv (mm)

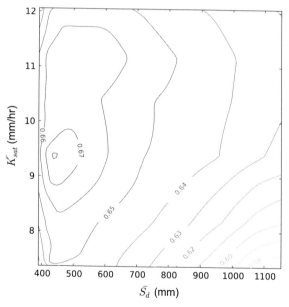

β_0n. Figures 10.4 and 10.5 show the sensitivity of the hydrologic response to hydraulic conductivity (both volume and peak increase with decreasing scalar, α) and hydraulic roughness (only the peak increases with decreasing scalar, β).

The response in hydrograph shape and magnitude produced by scalar adjustment is the basis for adjusting the simulated hydrograph to match observed ones. These

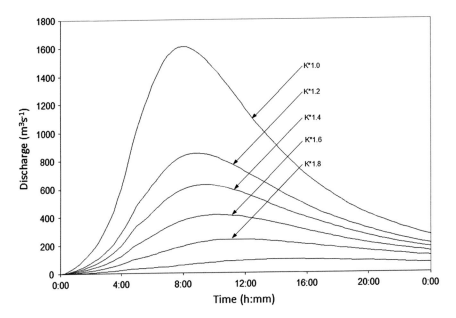

Fig. 10.4 Effect of adjusting hydraulic conductivity scalars on hydrograph peak and volume response

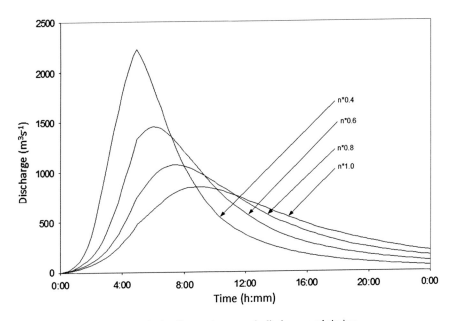

Fig. 10.5 Effect of adjusting hydraulic roughness peak discharge and timing

two parameters are the most important for this type of runoff model. Adjustment of parameter maps should sequentially be applied to drainage areas between gauging stations. Another refinement involves adjustment of roughness applied differentially to overland and channel roughness. Additional details on the OPPA method can be found in Vieux and Moreda (2003). Application of the OPPA method was used in both the DMIP I and II experiments (Smith et al. 2004, 2012; Vieux et al. 2004) and in urban flood modeling presented by Looper et al. (2012).

10.4 Parameter Search Algorithms

A principal objective of Data Assimilation methods is the retrieval of the state in atmospheric or hydrologic models (Le Dimet et al. 2009; Sene 2012 p. 125). Considering the many degrees of freedom in a distributed model, some have suggested that calibration is not possible. However, Vieux et al. (1998b) and White et al. (2002, 2003) showed that an identifiable optimum can be identified using an adjoint method. McLaughlin (1995) observed that there are few examples of distributed watershed models that assimilate field data. This may be related to model complexity, lack of observations, nonlinearity or thresholds (CRR bucket models) that are not compatible with the more widely used data assimilation methods in meteorology and oceanography. When the governing equations are not partial differential equations, as in the case of CRR models, inverse methods are precluded. Other methods of automatic calibration, such as the Kalman filter (Kalman and Bucy 1961), have been applied for state retrieval and to a limited extent in model calibration (Hosseinpour et al. 2014). One limitation is that the Kalman filter is applicable to linear problems and diverges in the case of nonlinearity or with large model errors (Entekhabi et al. 1994; McLaughlin 1995; Graham and McLaughlin 1991). Sene (2012), Blum et al. (2009), and Le Dimet et al. (2009) described data assimilation in hydrometeorological forecasting.

As hydrologists advance from empirical models toward PBD models that solve conservation equations, new methods of parameterization and calibration are needed, especially those that rely on the physics of the solution to guide parameter search and optimization. Adjoint methods and optimal control techniques are expected to become more common place as PBD models enter operational forecasting environments. The WRF-HYDRO model implemented by the NWS is a PBD model solving the DWA equations with *nudging,* a type of DA, to help enforce agreement between modeled and observed streamflow (Yucel et al. 2015; Cosgrove 2015).

Identification of optimal parameters that minimize a cost function composed of observed and simulated streamflow forms the inverse problem. The adjoint model is the inverse of the governing differential equation in the presence of data and it is constrained by an optimality condition (Le Dimet et al. 1996; 2009). Vieux et al. (1998a, b) used the adjoint model to compute the gradient of the cost function with respect to each of the multipliers described above. Then an iterative gradient-based optimization algorithm was used to find the optimal parameters, which satisfy the optimality constraint.

The spatial pattern of parameters affecting infiltration or hydraulic roughness preconditions the search for optimal calibration parameters. If each grid cell can be grouped and adjusted with a multiplicative factor, then the degrees of freedom are reduced. Put in terms of an optimal control framework, the inverse model is formed from the linear tangential and adjoint models, which allows computation of optimal calibration parameters and model sensitivity. Whether the problem is ill-conditioned and whether non-unique sets of parameters exist, it is characterized by the cost function surface.

10.4.1 Forward Model

For application to surface runoff and calibration, the KWA equation is the *forward or direct model*. Writing the KWA equation with scalars that multiply the parameter vectors results in

$$\frac{\partial h}{\partial t} + \frac{s^{1/2}}{\beta n}\frac{\partial h^{5/3}}{\partial x} = \gamma R - \alpha I \tag{10.12}$$

where the three scalars α, γ, and β adjust the multipliers controlling infiltration rate, I, rainfall rate, R, and hydraulic roughness, n, respectively; h is the flow depth and s is the principal land surface slope at the center of each grid cell. The slope and hydraulic roughness are spatially variable, while rainfall, infiltration, and flow depth are spatially and temporally variable. If we consider the rainfall as perfectly known, with no bias, then γ is equal to one. However, the problem may be cast in terms of a mean field bias adjustment of radar rainfall estimates.

10.4.2 Inverse Model

The adjoint model is the inverse of the forward model. In the application of optimal control theory, a model may be generally described by an equation of the following form:

$$\frac{\mathrm{d}X}{\mathrm{d}t} = F(X, K), \quad X \in \Re^n, \quad X(0) = X_0 \tag{10.13}$$

where X is a state variable, t is the time, K is a parameter, and X depends on K. The disagreement between the simulated value X and the corresponding observation X_{obs} could be quantified by the following cost function, J, which depends implicitly on K:

$$J(K) = \frac{1}{2} \int_0^\tau (X - X_{obs})^2 dt \qquad (10.14)$$

where J is minimum when the optimality condition $\nabla J(K^*) = 0$ is satisfied. The gradient of the cost function is obtained using the adjoint model, which needs the following definition of a function's directional derivative to be found:

$$\hat{X} = \lim_{l \to 0} \frac{X(K + lh) - X(K)}{l} \qquad (10.15)$$

where h is the directional derivative. This definition, applied to Eq. (10.12), gives rise to the tangential linear model (TLM) system:

$$\frac{d\hat{X}}{dt} = \left[\frac{\partial F}{\partial x}\right]\hat{X} + \left[\frac{\partial F}{\partial K}\right]h, \quad \hat{X}(0) = 0 \qquad (10.16)$$

The TLM (Eq. 10.16) is used to find the adjoint model. We now compute the directional derivative of the cost function as

$$\hat{J}(k, h) = \lim_{l \to 0} \frac{J(k + lh) - J(K)}{l} = \int_0^T ((X(K, t) - X_{obs}(t)\hat{X}(t))dt \qquad (10.17)$$

By introducing an adjoint variable, P, having the same dimension as X, the scalar product of P and the tangential linear system may be integrated by parts between 0 and τ. This integration calculates the adjoint variable, P, which is the solution of

$$\frac{dP}{dt} + \left[\frac{\partial F}{\partial x}\right]^T P = X - X_{obs} \qquad (10.18)$$
$$P(T) = 0$$

With an optimality condition, the directional derivative of the cost function can be written as

$$J = \langle h, -\int_0^\tau \left[\frac{\partial F}{\partial K}\right]^T P\, dt \rangle = \langle h, \nabla J \rangle \qquad (10.19)$$

Thus, the cost function gradient formula can be expressed as

$$\nabla J = -\int_0^\tau \left[\frac{\partial F}{\partial K}\right]^T P\, dt \qquad (10.20)$$

This cost function gradient can now be searched for parameters that minimize the difference between observed and simulated discharge values.

Because the forward model is discrete in time and space, we must form a discrete version of the inverse model, i.e., Eq. (10.18). As described in Chap. 9, the Galerkin formulation of the finite element method together with a finite difference scheme in time is applied to the KWA equations. The elemental equations are written in terms of linear shape functions to approximate the linear variation of the dependent variables across the element. A one-dimensional form suffices if the finite elements are laid out a priori in the direction of land surface slope. Then the time discretization uses the weighted Euler's method to compute the flow area A. This method gives the following system:

$$CA^{i+1} = CA^i + \Delta t \, F(\gamma R^i - \alpha I^i) - \beta \Delta t S Q^i \tag{10.21}$$

where Q is the discharge or flow rate; C and S (called, respectively, capacitance and stiffness matrices) depend on the topology of the finite elements connecting each grid cell in the river basin; and F is the forcing vector. Using the same method as in the theoretical continuous form, we can obtain the adjoint model with P as the adjoint variable:

$$
\begin{aligned}
P^{n-1}C &= A^n - A^n_{\text{obs}} \, C^T P^{i-1} - C^T P^i \\
&\quad + \frac{5\beta\Delta t}{3} \frac{s^{1/2}}{w^{2/3}n} \left[S\left(A^{2/3}\right)^i \right]^T P^i \\
&= A^i - A^i_{\text{obs}} \quad i = n-1, \ldots, 0
\end{aligned}
\tag{10.22}
$$

The gradients of the cost function with regard to each Lagrange multiplier, α, γ, and β, are written as

$$\nabla_\alpha J = -\Delta t \sum_{i=0}^{n-1} P^i F I^i \tag{10.23}$$

$$\nabla_\beta J = -\Delta t \sum_{i=0}^{n-1} P^i S Q^i \tag{10.24}$$

$$\nabla_\gamma J = \Delta t \sum_{i=0}^{n-1} P^i F R^i \tag{10.25}$$

By the adjoint method, the gradients with respect to each scalar multiplier are defined and used to search the parameter space to find the minimum of the cost surface. In practice, these quantities are computed numerically by the discrete inverse model at each time step. Applying the adjoint method involves the following steps:

1. Select an initial starting set of parameters.
2. Calculate the gradient of the cost function with regard to each Lagrange multiplier using the adjoint method.
3. Determine a search direction depending on the gradients obtained.
4. Find a new set of parameters in using the search direction calculated.
5. Check convergence criteria for termination. If not satisfied, return to Step 1 with the new set of parameters.

The adjoint method provides the gradients (Eqs. 10.23, 10.24, and 10.25) with respect to α, β, and γ in Step 1. However, in order to realize Steps 2, 3, and 4, the Broyden–Fletcher–Goldfarb–Shanno (BFGS) One-Step, Memoryless, Quasi-Newton Method is used (Liu and Nocedal 1989). This subroutine solves the unconstrained minimization problem. The advantage of the BFGS algorithm is that it does not require knowledge of g the Hessian matrix. Whether optimal parameters can be retrieved depends on the uniqueness of the solution, which is another way of stating that the differential equations constrained by observations are invertible and an optimum parameter set exists.

10.5 Calibration Example

The Flint River is a 280 km^2 subbasin of the Illinois River in Oklahoma, which was a subject basin in DMIP I and II. The optimal control technique is tested by *generating* an 'observed' hydrograph and then feeding this hydrograph back into see if it can retrieve the parameters used to generate the 'observed' hydrograph. Assimilating the 'observed' hydrograph into the inverse model yields the scalar multipliers used to create the hydrograph. A series of hourly maps of rainfall, produced by the WSR-88D radar, are input into the forward model to simulate the hydrograph shown on the left in Fig. 10.6. The cost function in Fig. 10.7 is produced given the 'observed' hydrograph. The parameter space consists in the

Fig. 10.6 Simulated and observed hydrographs generated at the outlet. *Inset* Resulting hydrograph obtained by retrieving optimal parameters

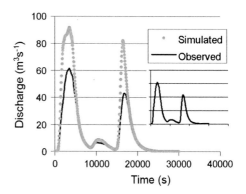

Fig. 10.7 Cost surface
showing single minimum of
the cost function

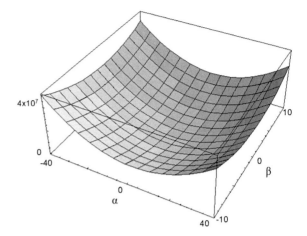

hydraulic conductivity multiplier, α, and hydraulic roughness multiplier, β. The vertical access is the cost function, J_v.

The scalar multipliers tested produce a well-conditioned cost surface on which a single-optimum value can be found. The forward model is invertible and a unique solution does exist. The smooth shape of the cost surface differs from the cost surfaces found for infiltration parameters assimilated from actual data in Figs. 10.2 and 10.3.

10.6   Summary

Physics-based distributed hydrologic modeling differs from conceptual or empirically based models because conservation laws are differential equations used to generate hydrologic quantities. Consequently, the model response to parameter adjustment and interactions between parameters behave in a predictable manner. Optimal values may be identified within suitable constraints, i.e., the spatial pattern of the parameter. If this spatial pattern is obtained from GIS/RS data, then we can adjust it with scalars to achieve agreement between observed and simulated hydrographs. The PBD calibration strategy consists in using multipliers to adjust the magnitude of the spatially distributed parameter while preserving its spatial pattern. Multipliers that reduce the difference between observed and simulated hydrographs results in maps of parameters considered optimal. Finding optimal parameters through application of multipliers reduces the dimension of the calibration problem. While the underpinnings are highly mathematical, the fact that an optimal parameter set can be shown to exist for a PBD model is a major advance in our understanding of how to apply distributed models to real-world hydrology.

References

Blum, J., F.-X. Le Dimet, and I.M. Navon. 2009. Data assimilation for geophysical fluids. *Handbook of numerical analysis* 14: 385–441.

Boyle, D.P., H.V. Gupta, S. Sorooshian, V. Koren, Z. Zhang, and M. Smith. 2001. Toward improved streamflow forecasts: Value of semidistributed modeling. *Water Resources Research* 37(11): 2749–2759.

Cheng, C.T., C.P. Ou, and K.W. Chau. 2002. Combining a fuzzy optimal model with a genetic algorithm to solve multi-objective rainfall–runoff model calibration. *Journal of Hydrology* 268(1): 72–86.

Cosgrove, B. 2015. December. Hydrologic modeling at the national water center: Operational implementation of the WRF-Hydro model to support national weather service hydrology. In *2015 AGU fall meeting*. AGU.

Duan, Q., V.K. Gupta, and S. Sorooshian. 1993. A shuffled complex evolution approach for effective and efficient optimization. *Journal of Optimization Theory Application* 76(3): 501–521.

Duan, Q., S. Sorooshian, and V.K. Gupta. 1992. Effective and efficient global optimization for conceptual rainfall-runoff models. *Water Resources Research* 24(8): 1015–1031.

Duan, Q., S. Sorooshian, and V.K. Gupta. 1994. Optimal use of the SCE-UA global optimization method for calibrating watershed models. *Journal of Hydrology* 158: 265–284.

Duan, Q., S. Sorooshian, H.V. Gupta, A.N. Rousseau, R. Turcotte, 2003. *Calibration of watershed models, water science and application series*, 6. American Geophysical Union, ISBN 0-87590-355-X.

Entekhabi, D., H. Nakamura, and E.G. Njoku. 1994. Solving the inverse problem for soil moisture and temperature profiles by sequential assimilation of multi-frequency remotely sensed observations. *IEEE Transactions on Geoscience and Remote Sensing* 32(2): 438–448.

Francés, F., Vélez, J.I., and Vélez, J.J., 2007. Split-parameter structure for the automatic calibration of distributed hydrological models. *Journal of Hydrology* 332(1): 226–240.

Graham, W.D., and D.B. McLaughlin. 1991. A stochastic model of solute transport in groundwater: Application to the Borden, Ontario, tracer test. *Water Resources Research* 27(6): 1345–1359.

Gupta, H.V., K.J. Beven, and T. Wagener. 2005. Model calibration and uncertainty estimation. *Encyclopedia of Hydrological Sciences*.

Hooke, R., and T.A. Jeeves. 1961. Direct search solutions of numerical and statistical problems. *Journal of Association for Computing Machinery* 8(2): 212–229.

Hosseinpour, A., Dolcine, L., and Fuamba, M., 2014. Natural flow reconstruction using kalman filter and water balance–based methods I: Theory. *Journal of Hydrologic Engineering* 19(12): 04014029.

Kalman, R.E., and R.S. Bucy. 1961. New results in linear filtering and prediction theory. *Journal of Basic Engineering* 83(1): 95–108.

Koren, V.I., M. Smith, D. Wang, and Z. Zhang. 2000. Use of soil property data in the derivation of conceptual rainfall runoff model parameters, paper presented at the 15th Conference on Hydrology. *American Meteorological Society Long Beach, California*.

Le Dimet, F.X., H.E. Ngodock, and M. Navon. 1996. Sensitivity analysis in variational data assimilation. In *Siam meeting in automatic differentiation*. New Mexico, USA: Santa Fe.

Le Dimet, François-Xavier, et al. 2009. Data assimilation in hydrology: variational approach. In *Data assimilation for atmospheric, oceanic and hydrologic applications*, 367–405. Berlin Heidelberg: Springer.

Liu, D.C., and J. Nocedal. 1989. On the Limited BFGS method for large scale optimization. *Journal of Mathematical Programming* 45: 503–528.

Looper, J.P., B.E. Vieux, and M.A. Moreno. 2012. Assessing the impacts of precipitation bias on distributed hydrologic model calibration and prediction accuracy. *Journal of Hydrology* 418–419: 110–122.

McLaughlin, D. 1995. Recent developments in hydrology data assimilation. *Review of Geophysics* 977–984.

Mendoza, P.A., M.P. Clark, M. Barlage, B. Rajagopalan, L. Samaniego, G. Abramowitz, and H. Gupta. 2015. Are we unnecessarily constraining the agility of complex process-based models? *Water Resources Research* 51: 716–728. doi:10.1002/2014WR015820.

Nash, J.E., and J. Sutcliffe. 1970. River flow forecasting through conceptual models, Part IA discussion of principles. *Journal of Hydrology* 10: 282–290.

Nelder, J.A., and R. Mead. 1965. A simplex method for function minimization. *Computer Journal* 7: 308–313.

Niedda, M., 2004. Upscaling hydraulic conductivity by means of entropy of terrain curvature representation. *Water resources research* 40(4).

Oudin, L., Andréassian, V., Perrin, C., and Anctil, F., 2004. Locating the sources of low-pass behavior within rainfall-runoff models. *Water Resources Research* 40(11).

Oudin, L., C. Perrin, T. Mathevet, V. Andreassian, and C. Michel. 2006. Impact of biased and randomly corrupted inputs on the efficiency and the parameters of watershed models. *Journal of Hydrology* 320(1): 62–83.

Pokhrel, P. (2007). Estimation of spatially distributed model parameters using a regularization approach.

Pokhrel, P.H., V. Gupta, and T. Wagener. 2008. A spatial regularization approach to parameter estimation for a distributed watershed model. *Water Resources Research* 44: W12419. doi:10.1029/2007WR006615.

Pokhrel, P.H., and V. Gupta. 2010. On the use of spatial regularization strategies to improve calibration of distributed watershed models. *Water Resources Research* 46: W01505. doi:10.1029/2009WR008066.

Sene, Kevin. 2012. *Flash floods: Forecasting and warning.* Springer Science & Business Media.

Smith, M.B., D.J. Seo, V.I. Koren, S. Reed, Z. Zhang, Q.Y. Duan, S. Cong, F. Moreda, and R. Anderson. 2004. The distributed model intercomparison project (DMIP): Motivation and experiment design. *Journal of Hydrology* 298(1–4): 4–26.

Smith, M.B., V. Koren, Z. Zhang, Y. Zhang, S.M. Reed, Z. Cui, F. Moreda, B.A. Cosgrove, N. Mizukami, E.A. Anderson, and DMIP 2 Participants. 2012. Results of the DMIP 2 Oklahoma experiments. *Journal of Hydrology* 418–419: 17-48.

Sorooshian, S., and J.A. Dracup. 1980. Stochastic parameter estimation procedures for hydrologic rainfall-runoff models: correlated and heteroscedastic error cases. *Water Resources Research* 29(4): 1185–1194.

Vieux, B.E., and N.S. Farajalla. 1994. Capturing the essential spatial variability in distributed hydrological modeling: Hydraulic roughness. *Journal of Hydrologic Processes* 8: 221–236.

Vieux, B.E., F. Le Dimet, D. Armand. 1998a. Optimal control and adjoint methods applied to distributed hydrologic model calibration. Proceedings of *International Association for Computational Mechanics*, IV World Congress on Computational Mechanics, June 29–July 2, Buenos Aires, Argentina. Computational Mechanics: New Trends and Applications, eds. S. R. Idelsohn, E. Onate, E.N. Dvorkin, CIMNE, Barcelona, Spain.

Vieux, B.E., F. Le Dimet, D. Armand. 1998b. Inverse problem formulation for spatially distributed river basin model calibration using the adjoint method. EGS, *Annales Geophysicae*, Part II, Hydrology, Oceans and Atmosphere, Supplement II to Vol. 16, p. C501.

Vieux, B.E., and F.G. Moreda. 2003. Ordered Physics-Based Parameter Adjustment of a Distributed Model. In: ed. Q. Duan, S. Sorooshian, H.V. Gupta, A.N. Rousseau, R. Turcotte, *Calibration of watershed models,* Water Science and Application Series, 6, 267–281, American Geophysical Union, ISBN 0-87590-355-X.

Vieux, B.E., and F.G. Moreda. 2003. Ordered physics-based parameter adjustment of a distributed model. In: Ed. Q. Duan, H.V. Gupta, S. Sorooshian, A.N. Rousseau, R Turcotte, *Calibration of watershed models, Water science and application,* vol. 6, 267–281. American Geophysical Union.

Vieux, B.E., Z. Cui, and A. Gaur. 2004. Evaluation of a physics-based distributed hydrologic model for flood forecasting. *Journal of Hydrology* 298(1): 155–177.

Vieux, B.E., Jin-Hyeog Park, and Boosik Kang. 2009. Distributed hydrologic prediction: Sensitivity to accuracy of initial soil moisture conditions and radar rainfall input. *Journal of Hydrology* 14(7): 671–689.

White, L.W., B.E. Vieux, and D. Armand. 2002. Surface flow model: Inverse problems and predictions. *Journal Advances in Water Resources* 25(3): 317–324.

White, L., B. Vieux, D. Armand, and F.-X. Le Dimet. 2003. Estimation of optimal parameters for a surface hydrology model. *Journal Advances in Water Resources* 26(3): 337–348.

Wagener, T., H.S. Wheater, and H.V. Gupta. 2004. *Rainfall-runoff modelling in gauged and ungauged catchments*, 306. London: Imperial College Press.

Yapo, P.O., H.V. Gupta, and S. Sorooshian. 1996. Automatic calibration of conceptual rainfall-runoff models: Sensitivity to calibration data. *Journal of Hydrology* 181(1): 23–48.

Yapo, P.O., H.V. Gupta, and S. Sorooshian. 1998. Multi-objective global optimization for hydrologic models. *Journal of Hydrology* 204: 83–97.

Yucel, I., A. Onen, K.K. Yilmaz, and D.J. Gochis. 2015. Calibration and evaluation of a flood forecasting system: Utility of numerical weather prediction model, data assimilation and satellite-based rainfall. *Journal of Hydrology* 523: 49–66.

Chapter 11
Case Studies in Distributed Hydrology

Abstract Case studies are presented that illustrate the principles of physics-based distributed hydrologic modeling. This chapter is organized by case study with a general description of the model, parameterization, and simulation used in each case. Within the context of three case studies, there is a common theme, namely each case study relies on high-resolution rainfall derived from radar and rain gauge; model parameters developed using geospatial data; and hydrologic predictions generated for analyzes spanning a range of space-time scales.

11.1 Introduction

The distributed model, V$flo^{®}$, is used to provide flood forecasting for basins located in regions where flooding is prevalent. Because of steep terrain coupled with abundant tropical moisture from the Gulf of Mexico, central Texas is prone to floods, especially in urban and peri-urban watersheds. An example of a flood forecast is shown in Fig. 11.1. There were copious rainfall amounts, exceeding 355 mm during a 6-h period (see band of red pixels). Rainfall intensity in some GARR pixels exceeded 235 mm/h in a 15-min period. These broadly distributed rainfall depths and intensities measured by GARR and DPR in real-time, drove the observed and modeled hydrograph peak response to less than 6 h (see Fig. 11.1 on right). Both the GARR and DPR forecast hydrographs are in close agreement with the observed stage. From base flow to peak the rate of rise was 41 m/h. Details on model performance during a significant flood event in 2010, caused by Tropical Hermine and its remnants, arc described by Looper and Vieux (2012).

Distributed hydrologic modeling within the context of three case studies is presented below. Each case relies on distributed rainfall estimation and model parameters derived from geospatial data to produce hydrologic predictions across a range of space-time scales.

- Case Study I—Reservoir Inflow Forecasting
  Reservoir inflow is a key component necessary for efficient power generation and flood management at upstream locations along the impoundment and in

© Springer Science+Business Media Dordrecht 2016   211
B.E. Vieux, *Distributed Hydrologic Modeling Using GIS*,
Water Science and Technology Library 74, DOI 10.1007/978-94-024-0930-7_11

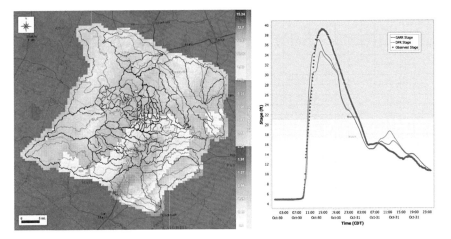

Fig. 11.1 Operational forecasting results during the 2015 'Halloween' flood in central, Texas. GARR rainfall accumulation (*on left*) is used to produce forecast hydrographs (*on right*) (35.3 cfs = 1 m^3 s^{-1}; 1 ft. = 0.3048 m)

downstream river reaches. A hydrometeorological network is used to measure current precipitation and numerical weather prediction for hydrologic predictions out to 7-days.

- Case Study II—Urban Flood Forecasting
  Distributed flood forecasting in an urban drainage context with several events reconstructed from archived radar rainfall. Operational flood forecasts in an urban basin are presented for a recent flood event.
- Case Study III—Climate Change Impact Assessment
  Estimation of climate change impacts on streamflow is accomplished with a calibrated model (1995–2013) of the Canadian River basin. Streamflow impacts are projected at ecological sampling sites by perturbing precipitation and PET derived from GCM projections.

11.2 Case Study I—Reservoir Inflow Forecasting

Operating a hydroelectric dam today is more difficult than ever considering business requirements for power generation, respecting downstream environmental releases, as well as management of lake levels that affect property owners along the lakeshore. Such operating constraints increase the need for accurate inflow forecasts. A reservoir is located in Central Missouri on the Osage River. The lake's serpentine shape, which terminates at a dam is seen in Fig. 11.2. The main channel is 94 miles in length from end to end and requires specialized cells for routing flow through shallow water within the body of the reservoir. Upstream releases from a flood

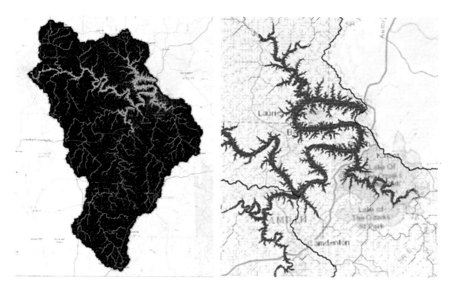

Fig. 11.2 Lake of the Ozarks tributary drainage area downstream of HST including the Niangua River (*left*) and a zoom over the lake and dam (*right*)

control reservoir contribute inflow along with runoff from an uncontrolled drainage area in the Niangua River comprising 6,241 km^2 (2,410 mi^2). The V*flo*$^®$ model of the uncontrolled drainage consists in 25,458 grid cells at 500-m resolution.

How much the reservoir level will change in hours, days, or weeks is important for managing power generation, especially when a lake level rise or downstream discharge imposes limits. A continuous physics-based distributed model is used with advantage to provide knowledge of reservoir inflow from diffuse streams, rivers and the discharge from an upstream reservoir. The physics-based approach is used to provide reservoir inflow forecasts with input from current and forecast rainfall, uncontrolled watershed runoff, and current or planned releases from the upstream reservoir (Vieux et al. 2015).

Predicted or forecast reservoir inflow aids in the operation of the dam, provides estimated spill consistent with its operating license and supports the forecast of lake levels with quantitative precipitation forecasts (QPF) from numerical weather prediction (NWP). V*flo*$^®$ is at the core of the software system that integrates precipitation and evapotranspiration data for making inflow projections, supporting web-based delivery of hydrologic information across a range of short (6-h) to longer range (7-day) time frames.

Streamflow is simulated within a drainage network composed of overland, channel, and reservoir body cells that route runoff to the dam. Gate operations are transmitted to the model in real time so that the lake level and outflow are considered in the modeling system. Soil moisture and infiltration rates are simulated in each grid cell using Green and Ampt parameters for a single layer and specified depth. Thus, both Horton and Dunne runoff components are computed, allowing the infiltration rate and saturation excess to operate simultaneously in the watershed. Soil moisture is depleted at hourly rates by PET derived from Reference ETo computed from measured humidity, wind speed, solar radiation, pressure, and temperature (Walter et al. 2000). Forecast PET is ingested from the National Digital Forecast Database, while climatological PET (Farnsworth et al. 1982) serves as a fallback if current measurements or forecast PET are unavailable in operations. Operation of the dam requires continuous simulation for a range of runoff events from small to large. To meet this objective, the model is setup and calibrated for a range of events within a continuous record.

In order to create a detailed model of the watershed capable of modeling the often highly variable response to soil moisture and rainfall distributions, a gridded model is setup that leverages geospatial data available across the US. Geospatial data availability means distributed models can incorporate information from maps of land use or vegetative cover, complex terrain, and soils to make the needed predictions of rainfall-runoff. When rainfall derived from weather radar and rain gauges is properly controlled for accuracy, it can be used as a necessary ingredient for operational modeling of the watershed. Model calibration was used to fine tune the model to be more accurate using historical rainfall data. Evaluation of the model performance from an operational review provides insight into how the system can be improved when there is a trend in over- or under-prediction. The information system that manages the rainfall production, modeling and data/information display, and interconnections is tailored to provide specific requirements for forecasting reservoir inflows. An example of a forecast lake level based on current and forecast rainfall is presented in Fig. 11.3 for a high flow period in December, 2015.

11.2.1   Hydrometeorological System

For a rainfall-runoff model to yield useful inflow forecasts, accurate precipitation quantification is necessary. Rather than rely on uncertain estimates from the NWS MPE, which is only available from the NWS at hourly and 4 × 4 km resolution, a higher resolution product is operated, which is customized to the target watershed. Three radars are bias corrected and merged to produce GARR at 15-min and 2 × 2 km resolution for model input. Automated failover from primary to secondary radars and then to gauge-only gridded rainfall occurs in the unlikely event that all three radars become unavailable. The hydrometeorological setup in Fig. 11.4 shows the locations of the three radars used to provide GARR over the tributary area on the Osage and Niangua Rivers. Data are also collected in near real-time from 135 precipitation stations obtained from federal and local agency sources and used in bias correction of the radar rainfall estimates.

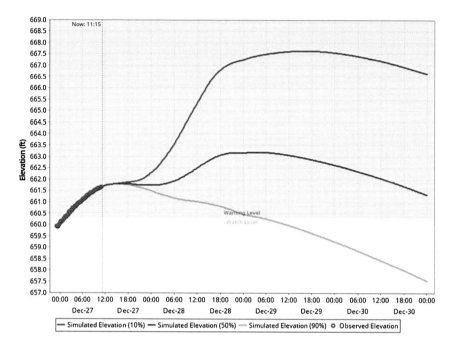

Fig. 11.3 Lake level forecasts based on current and probabilistic precipitation forecast inputs (1 ft. = 0.3048 m)

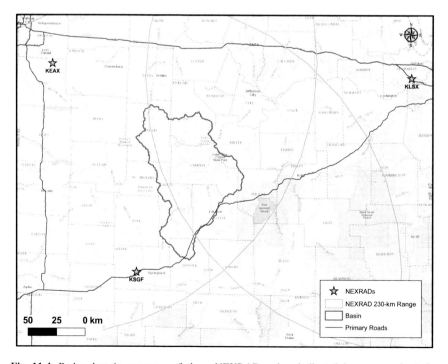

Fig. 11.4 Radar domain coverage of three NEXRAD radars indicated by *stars* and *circles* specifying a range of 230-km from each station

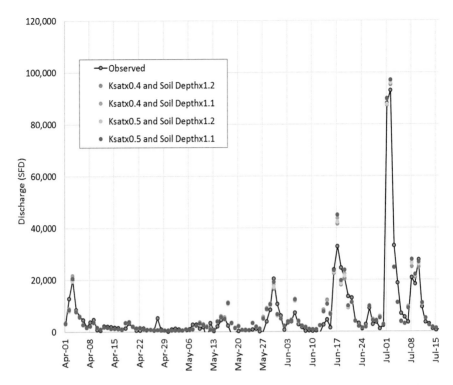

Fig. 11.5 Continuous model calibration used to identify optimal parameters for a period of high flow on July 1–2, 2015 (1 s foot day (SFD) = 35.31 cfs day = 1 m³ s⁻¹ day)

11.2.2   Model Parameter Adjustment

Continuous simulation with measured ETo results in optimization of the model where adjustments to saturated hydraulic conductivity, K_{sat} and soil depth, S_d, help achieve minimization of the Nash Sutcliffe and mean absolute percentage error (MAPE). Figure 11.5 shows the model response for a range of multipliers applied to distributed maps of each parameter, namely, K_{sat} and S_d. For continuous simulation starting January 1, optimal results are achieved with multipliers of $K_{sat} \times 0.3$ and $S_d \times 1.3$. The result is that $K_{sat} = 1.09$ cm/h is reduced to 0.43 cm/h and S_d increases from 63.5 to 82.6 cm. PET can be formulated by several methods including climatological values or measured atmospheric parameters within the ASCE reference ET method (Walter et al. 2000).

In Fig. 11.6, the three PET inputs tested are shown: Run (1) measured PET for 2015, Run (2) climatological PET and Run (3) operational climatological PET with crop coefficients.

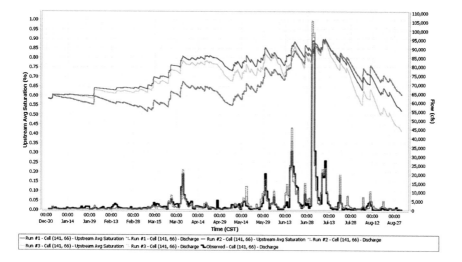

Fig. 11.6 Sensitivity to PET by continuous simulation during Dec 30–Aug 31, 2015 ($35.31 \text{ cfs} = 1 \text{ m}^3 \text{ s}^{-1}$)

11.3 Case Study II—Urban Flood Forecasting

Floods have plagued the Houston, Texas metropolitan area for many decades. Major floods occurred in the greater Houston metropolitan area in 1983, 1989, 1992, 1994, 1996, 1998, and 2001. Deployment of physics-based distributed models for operational flood forecasting is now in practice (Bedient et al. 2003; Vieux et al. 2003, 2015; Vieux and Bedient 2004; Looper et al. 2012). Operational flood forecasting in urban areas differs in terms of scale and purpose from those systems supporting national-level flood forecasting responsibilities. Site-specific forecasts in urban areas require customized flood forecasting systems rather than the more generalized warnings produced by the NWS/NWM or the NWS/River Forecast Centers (RFCs). A customized operational flood forecasting system provides site-specific information to a regional medical center located in Houston, Texas. This system supports operations and logistical measures to reduce flood losses in the hospital complex and related facilities. Because of its location in an urbanized watershed, the medical center is vulnerable to flooding whenever sufficiently intense and prolonged rainfall occurs upstream. When flooding is imminent in Brays Bayou, which flows adjacent to the medical center, specific actions are taken. These include placing member institutions on alert, closing of floodgates and doors, or suspending patient care and evacuating the hospitals/facilities. The forecast point of interest is adjacent to the institution in Brays Bayou, where there happens to be a stream gauge operated by USGS.

During a major flood event, Tropical Storm (TS) Allison in June 2001, there was a suspension of services during a hospital shutdown. Successful forecasts resulted

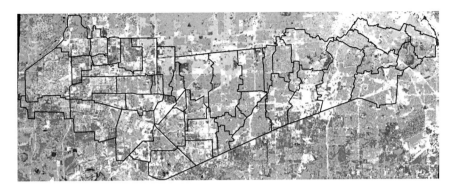

Fig. 11.7 Brays Bayou watershed subbasins and USGS gauging stations. The *shading* represents the overland hydraulic roughness parameter derived from Landsat 7 land use/cover classification

in continuity of services during TS Frances in September 1998 and again in November, 2003. Further details on this system and operational features may be found in Bedient et al. (2003), Vieux and Bedient (2004), Fang et al. (2011). The Brays Bayou watershed outline and gauging stations are shown in Fig. 11.7. Upstream of the USGS Main Street gauge, there are 16,717 grid cells at 120-m resolution. The shaded image depicts the hydraulic roughness parameter derived from reclassification of land use/land cover. Imperviousness is also incorporated into the model. The lighter shades of gray also indicate higher imperviousness at nearly 100 % in some portions of the subbasins.

As with any modeling system, performance depends on three categories of uncertainties: (1) rainfall input uncertainty; (2) model structure; (3) model parameters. Addressing the first, rainfall uncertainty is reduced through automated gauge correction for bias in the radar QPE. NEXRAD reflectivity (Level II) is transformed into rainfall rate by either the tropical relationship $Z = 250R^{1.2}$Tropical Z-R relationship or the convective relationship: $Z = 300R^{1.4}$. The tropical Z-R relationship developed by Rosenfeld et al. (1993) is used extensively in coastal areas of the US impacted by tropical storms and warm-process rainfall events. However, assumed Z-R relationships are only a starting point for operational bias adjustment. Once reflectivity is transformed into rainfall rates, they are compared to rain gauges in and around Brays Bayou and used to compute a local bias adjustment with excluded outliers as quality control. A plot of the bias correction factors produced by the operational system during a recent event is shown in Fig. 11.8.

11.3.1   Basin Characteristics

Brays Bayou has a drainage area of 260 km^2 that drains through a largely urbanized area. Three stream gauges are operated by the United States Geological Survey (USGS). These gauges are located in the basin at Main Street (USGS 08075000),

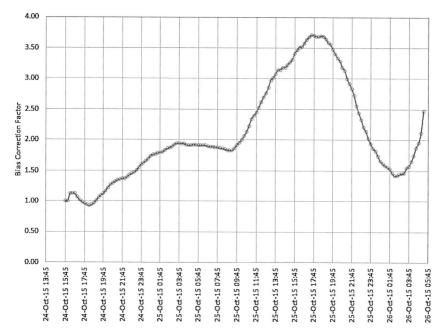

Fig. 11.8 Operational bias correction factors applied to radar during a recent flood event, October 24–26, 2015

Gessner and Roark Road. Manning hydraulic roughness values are assigned based on a 30-m Landsat 7 Thematic Mapper, Anderson classification of dominant land use categories including imperviousness. The watershed and surrounding region is highly urbanized with about 85 % of the watershed developed. The lower 42 km of the channel was concrete lined in the 1960s by the US Army Corps of Engineers. The concrete trapezoidal cross-section has a 15-m bottom width and 3:1 side slopes near Main Street. Extending to the headwaters, channel bottom widths decrease to ~5 m with the same 3:1 side slopes. Slopes in the overland and channel areas are quite flat with channel slopes of 0.001 % downstream of Main Street to the East. Channel slopes above Main Street are generally 0.055 % with upstream channel slopes in the headwaters around 0.2 %. The coastal soils are typically composed of clay with low infiltration rates. Soil conditions and impervious areas result in large fractions of rainfall being transformed into runoff. Hydraulic conductivity of non-pervious areas is based on soil properties and is generally assigned a value of 0.076 cm/h based on soil texture found in this basin, which is a clayey texture. Juan et al. (2015) found similar values of hydraulic conductivity for a basin located to the west of Brays, in Sugarland TX through calibration. They found somewhat higher values that ranged from 0.27 to 0.57 cm/h.

Parameter adjustment results in identification of scalars that minimize the objective functions of volume and then timing and peak discharge. Finding a single set of parameters (scalar multipliers) that minimizes the objective functions across

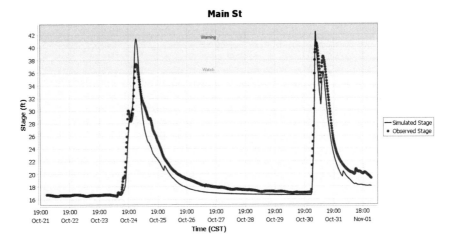

Fig. 11.9 Simulated and observed peak stage at Main Street October 23–November 01, 2015 (1 ft. = 0.3048 m)

several storms implies that the parameters are identifiable and stable. Verification using this parameter set at an interior point in the watershed provides confidence in the validity of the calibrated model. The OPPA procedure described by Vieux and Moreda (2003) follows these steps to adjust: (1) hydraulic conductivity to match volume; (2) overland hydraulic roughness adjustment to match timing and peak; and (3) channel hydraulic characteristics to improve time to peak. More details on the calibration of the model for a series of storms may be found in Vieux and Bedient (2004). Operation of the forecasting system for Brays Bayou was initiated in real time in 1997 and is still in use. Both the V$flo^{®}$ and HEC-1F models are operated continuously with GARR inputs (Bedient et al. 2003; Fang et al. 2011). Figure 11.9 shows the operational result with closely matching rising limbs where both simulated and observed outcomes exceed watch levels and nearly exceed the higher warning level.

11.4   Case Study III—Climate Change Impact Assessment

Estimation of climate change impacts on streamflow is accomplished with a calibrated model (1995–2013) of the Canadian River basin. Streamflow impacts are evaluated at ecological sampling sites by perturbing precipitation and PET. V$flo^{®}$ is setup and calibrated to a historic period of rainfall, namely 1995–2013, during which there are rain gauge and radar observations of precipitation and measured streamflow within the Canadian River basin. The resulting model is used to produce baseline discharge with derivative flow metrics at distributed interior basin locations where sampling exists for fish species. Projections of discharge under climate scenarios are made using a delta method, where precipitation and PET are perturbed by factors derived from GCM projections. The modeling approach relies on spatially distributed parameter maps governing runoff and maps of hourly precipitation input derived from a combination of weather radar and rain gauges for a period when these data were available, 1994–2013.

11.4.1   Precipitation Inputs

Precipitation measurements derived from radar and rain gauge measurements are referred to as Multisensor Precipitation Estimates (MPE). Hydrologic prediction requires precipitation that is both representative (spatially distributed) and accurate (bias corrected) for a given watershed area. Radar-derived precipitation estimates, if not bias corrected, can lead to inaccurate streamflow predictions and miscalibration if not addressed (Oudin et al. 2006; Looper et al. 2012). Young et al. (2000) documented inconsistencies in operationally produced MPE, such as range dependent biases and anomalies associated with melting of frozen hydrometeors. Therefore, the MPE precipitation was subjected to quality control of rain gauges to remove residual bias, fill gaps in the MPE record where radar was missing, and addition of rain gauge data not available operationally. This reanalysis of MPE resulted in more complete and reliable precipitation (GARR). The Canadian River Basin comprising a total drainage area of 70,188 km^2 was modeled as upper and lower watersheds (20,408 km^2) separated by Lake Meredith in the Texas Panhandle. The lower Canadian watershed shown in Fig. 11.10 shows a plot of cumulative and hourly rainfall for a rain gauge located at El Reno, Oklahoma (identified as ELRE). The GARR is sampled from the MPE grid (ID 32015) overlying this gauge location. A storm total for the 20-year period, 1994–2013, is presented in Fig. 11.11. There is a strong climatological gradient of rainfall from West to East, ranging from 782 to 2,159 cm. This corresponds to mean annual precipitation totaling 39–108 cm, or a 69 cm increase across a distance of 500 km spanning the lower basin, which is a 10 mm increase per 73 km from West to East.

Significant improvement resulted in terms of the average difference (AD) between radar and rain gauge daily event totals, decreasing from 56.5 to 17.5 %, i.e., to within ±8.7 %. Similar results were found for the lower Canadian watershed, where GARR accuracy improved from 25.3 to 13.9 % through bias correction with 55 rain gauges.

11.4.2   Potential Evapotranspiration

Because the modeling approach requires continuous simulation of streamflow, PET is used to control moisture fluxes to the atmosphere from the soil-vegetation complex. Increased PET during warmer periods accelerates the drying of the soil

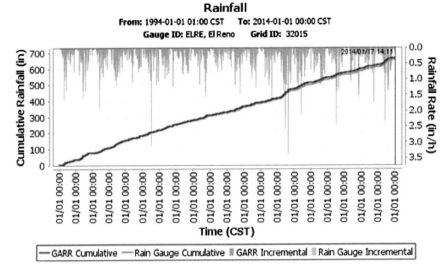

Fig. 11.10 Period of record cumulative and hourly rainfall for the El Reno (ELRE) rain gauge and GARR (1994–2013). 1 in. = 25.4 mm

profile and thereby reduces runoff. One approach is to apply climatological values rather than actual measurements, which are typically sparse, or only available for a portion of the historical period. Mean monthly PET rates were taken from CONUS map values at Oklahoma City, within the Lower Canadian (Farnsworth et al. 1982). In application of the model, the PET is modified by two processes; first, when the evaporative flux ceases because the soil moisture is depleted and second, when precipitation is occurring as determined by the overlying radar grid. In Fig. 11.12, the climatological mean monthly PET was assumed for the historical period and for future periods with perturbations applied as discussed below.

11.4.3 Hydrologic Modeling 'As-If' Climate Has Changed

Investigation of how the Canadian River basin would respond to changes in precipitation and PET relies on a physics-based distributed (PBD) hydrologic model. The background and development of the $V\!flo^{®}$ model is described in Vieux (2004); its application to continuous simulation using long-term archival GARR for computation of water balance components was presented by Moreno and Vieux (2013); comparison of PBD models with conceptual rainfall-runoff watershed models, including $V\!flo^{®}$, is contained in the special issue Smith et al. (2012); and sensitivity to reanalysis of MPE was evaluated by Looper et al. (2012). PBD modeling solves conservation equation of mass and momentum on discrete elements and in this case, a gridded representation of the land surface. Numerical solution of the kinematic

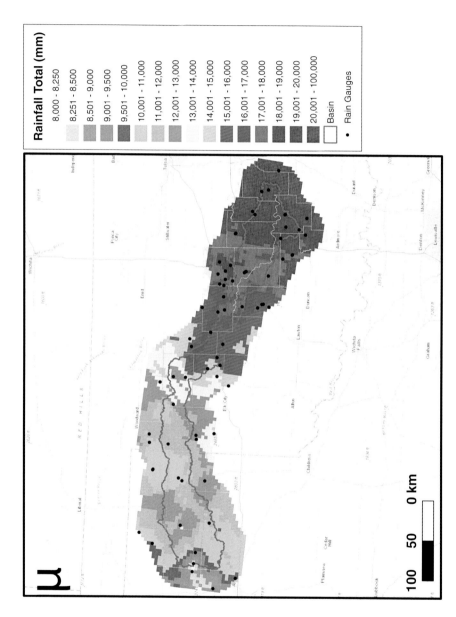

Fig. 11.11 GARR precipitation total for the historical period (1994–2013). 1 in. = 25.4 mm

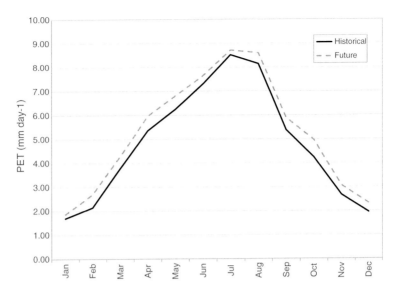

Fig. 11.12 Climatological monthly PET for a historical period and future projection

wave equation calculates discharge and depth of flow in each overland and channel grid cell comprising the drainage network. Watershed parameters are derived a priori from geospatial maps; namely, gridded terrain for slope and drainage network connectivity, soils maps for infiltration parameters, and land use/cover for hydraulic roughness. Model calibration at USGS stream gauging sites is performed to optimize the a priori model parameter maps with an objective function composed of simulated and observed runoff volume. Validation was performed for periods not used for calibration. Because flow metrics are needed at fish sampling sites, the calibrated model produces discharge in channel grid cells. Fish sampling sites are located throughout the main channel reach of the Canadian River but only a few sample sites are located in close proximity to USGS stream gauge sites. The Upper Canadian River model is composed of 51,596 cells at 1 × 1 km resolution. The Lower Canadian is composed of 19,207 cells also at 1 × 1 km resolution. Figures 11.13 and 11.14 illustrate the spatial distribution of sampling sites relative to USGS stream gauges in the Upper and Lower Canadian River, respectively.

11.4.4 Historic Period Calibration

To establish baseline runoff response, the Upper and Lower Canadian River models were calibrated using the historic period GARR (1994–2013) and climatological PET was considered to be representative of the historic period. The USGS stream gauges and the drainage area used for calibration are presented in Table 11.1 and

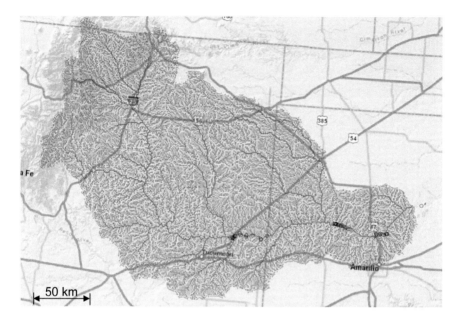

Fig. 11.13 Upper Canadian River model with sampling sites (*black circle icons*) superposed on the drainage network with overland cells (*gray*), channel cells (*blue*) and two collocated USGS stream gauges (*red stars*)

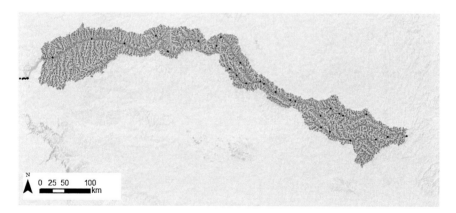

Fig. 11.14 Lower Canadian River model with sampling sites (*black circle icons*) superposed on the drainage network with overland cells (*gray*) and channel cells (*blue*)

Fig. 11.15. The most sensitive parameters controlling runoff are the Green and Ampt soil infiltration parameters, saturated hydraulic conductivity controlling infiltration rate excess runoff and soil depth that controls saturation excess runoff. Both processes can operate in the watershed, where infiltration rate excess is

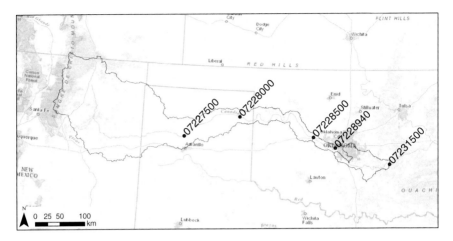

Fig. 11.15 Locations of USGS stream gauges used in calibration for the Upper and Lower Canadian River models

produced by high intensity rainfall and saturation excess generates runoff from long duration but lower intensity rainfalls that saturate the soil profile (Vieux 2004, pp. 91–112). These parameters were adjusted upstream of the five USGS stream gauges within the basin to achieve 1:1 agreement between cumulative runoff volumes plotted over the period of record.

11.4.5   Climate Change Perturbation

Assessment of climate change impacts is accomplished using the delta method (Hamlet and Lettenmaier 1999; Miller et al. 2003) where the analysis reflects a change in monthly mean temperature and precipitation. Because GCM models have resolutions that are too coarse for water resources response, bias correction and spatial downscaling (BCSD) are applied as described by Maurer (2007). Projections were selected from the BCSD GCM output as part of the World Climate Research

Table 11.1 USGS gauging stations used in model calibration for the historic period

Identifier	Location description	Area (km^2)	Area (mi^2)
7227500	Canadian River near Amarillo, TX	50,343	19,445
7228000	Canadian River near Canadian, TX	59,200	22,866
7228500	Canadian River at Bridgeport, OK	63,943	24,698
7228940	Canadian River near Mustang, OK	65,072	25,134
7231500	Canadian River at Calvin, OK	70,983	27,417

Programme (WCRP), Coupled Model Intercomparison Project, Phase 3 (CMIP3) (see IPCC 2007; Meehl et al. 2007). These projections were produced collectively by 16 different CMIP3 models simulating three emissions paths, B1 (low), A1b (middle) and A2 (high), from end of twentieth century climate conditions. The focus of this study was to select a model and its changes in monthly temperature and precipitation over the study region. Of the 112 projections, a preponderance of models showed drier and warmer conditions (Reclamation 2010). The ensemble hybrid approach for identifying projections was introduced and applied (Reclamation 2008, 2009). The ensemble member, INM CM3.0, Run 1 and Emission path, A2, exhibited a—10 % (decrease) in mean annual precipitation but a 90th percentile rank in terms of temperature increase. Monthly model projections for PET and Precipitation (1950–2099) are shown in Figs. 11.16 and 11.17, respectively. The 12-month moving average for each variable is displayed as a black line. PET is seen to increase and precipitation is decreasing while exhibiting considerable more variance. These GCM time series are used to develop monthly delta adjustments of temperature and precipitation. The past period was selected from the end of the twentieth Century, 1970–1999, which coincided with the historic GARR period. The future period at the end of the twenty-first century was selected as 2070–2099. Figures 11.18 and 11.19 show the precipitation and PET for the past and future periods used to perturb inputs from the historical period. The delta method was applied to the precipitation (GARR) and PET data to assess

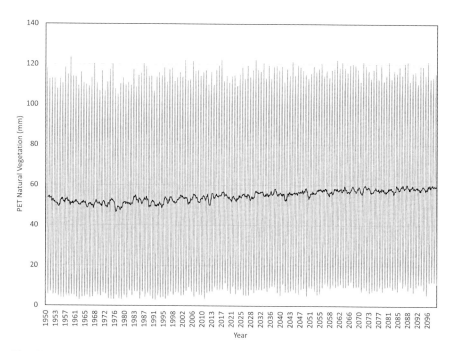

Fig. 11.16 Monthly PET for natural vegetation from 1950 to 2099, INM CM3.0

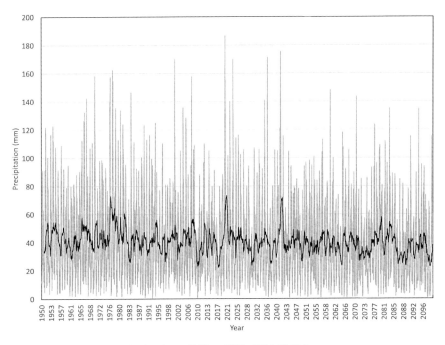

Fig. 11.17 Monthly precipitation from 1950 to 2099, INM CM3.0

Fig. 11.18 Mean monthly
precipitation projected by
INM CM3.0 for past (1970–
1999) and future (2070–2099)
periods

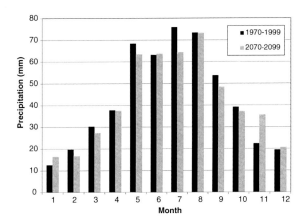

differences in streamflow between the past and future climate periods *as-if* the
climate has already changed (Vieux 2015).

Table 11.2 presents the ratios of precipitation of future/past precipitation.
Similarly, ratios of PET from the INM CM3.0 model were computed as seen in the last
column. The ratios for precipitation are generally less than 1.0 indicating drier conditions except for January, April, and November when the ratios are greater than 1.0

Fig. 11.19 Mean monthly
PET projected by INM
CM3.0 for past (1970–1999)
and future periods (2070–
2099)

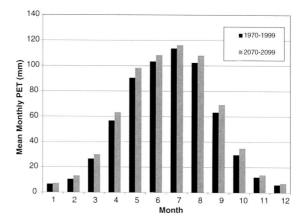

Table 11.2 Perturbation
ratios of past (1970–1999)
and future (2070–2099)
periods

Month	Precipitation	PET
1	1.14	1.10
2	0.58	1.26
3	0.95	1.13
4	1.07	1.11
5	0.82	1.09
6	0.78	1.05
7	0.81	1.02
8	0.88	1.06
9	0.71	1.09
10	0.70	1.17
11	1.21	1.15
12	0.96	1.19

(wetter). The warmer temperature profile is reflected by the PET ratios that are all
greater than 1.0 indicating a greater flux atmosphere in 2070–2099 than in 1970–1999.

The change in precipitation and PET projected for the future period is applied to
the GARR precipitation data from the historical period, 1994–2013, as-if climate
change has already occurred. The hydrologic response from the baseline to the
perturbed PET and precipitation inputs is used for evaluation of future climate
impacts. This method incorporates GCM-predicted changes in precipitation and
PET while preserving realistic spatial, temporal, and convective patterns of rainfall
typical of the region. While decreasing precipitation will likely affect base flow
quantities, it is not perturbed between baseline and future periods. More intense
rainfall is not evaluated even though a warmer atmosphere would likely lead to
more intense events in spite of the fact that over all precipitation is expected to
decrease under the warmer–drier climate projection. Further, land use or vegetation
changes induced by a warmer and drier climate were not modified when making

future hydrologic projections. Separate precipitation perturbations representing change in precipitation, in PET and in both precipitation and PET are evaluated to identify the relative magnitudes of hydrologic response induced by a changing climate.

11.4.6   Hydrologic Analysis

The calibrated V$flo^{®}$ model used to produce the baseline hydrologic response is then used to simulate future conditions. Continuous streamflow estimates are generated from the calibrated model for ungauged locations dictated by existing locations, water quality, and fish habitat sample data. Streamflow produced under future climate change scenarios are then used to derive streamflow metrics assessed for the two periods. These metrics are then used to make projections of how the ecosystem and fish populations may respond under the climate change. Figure 11.20 shows the period of record calibrated model streamflow, both simulated and observed for the historic period, 1994–2013, i.e., with historic GARR and PET. Figure 11.21 is the period of record hydrograph of baseline (simulated time series from Fig. 11.20) and perturbed hourly streamflow based on future precipitation and streamflow. Not shown are results from only perturbed precipitation keeping PET at historic values, which are nearly identical. Figure 11.22 shows the cumulative plots of baseline and future streamflow, which exhibit a marked and consistent decline over the 30-year simulated period at the end of the twenty-first century. In Fig. 11.23 the climatology of the baseline and future streamflow is shown. An interesting feature is the seasonal shift in streamflow from May to December; there is a clear diminution of monthly streamflow, which has potential ecological impacts that are discussed in the following sections.

Fig. 11.20 Calibrated model observed and simulated streamflow (1994–2013) ($35.31 \text{ cfs} = 1 \text{ m}^3 \text{ s}^{-1}$)

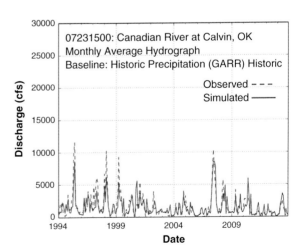

Fig. 11.21 Streamflow with both future precipitation and future PET
$(35.31 \text{ cfs} = 1 \text{ m}^3 \text{ s}^{-1})$

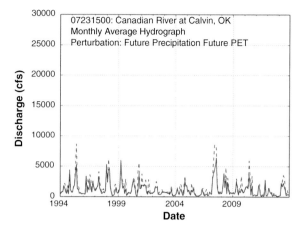

Fig. 11.22 Cumulative runoff for future precipitation and future PET
$(1 \text{ in.} = 25.4 \text{ mm})$

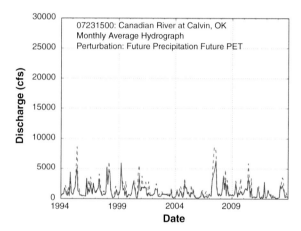

Fig. 11.23 Streamflow climatology for baseline and perturbed (future precipitation and future PET) $(35.31 \text{ cfs} = 1 \text{ m}^3 \text{ s}^{-1})$

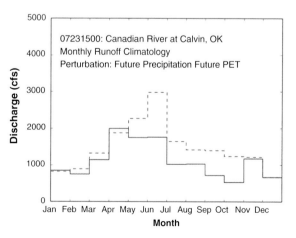

11.5 Summary

This chapter presented three case studies that illustrate factors affecting distributed modeling. They also demonstrate the capabilities of a distributed model for both rural and urban application areas. Physics-based models use conservation of mass and momentum, referred to as a hydrodynamic or hydraulic approach to hydrology. The opening Fig. 11.1 illustrates a central concept of this book, i.e., the transformation of radar-based precipitation into rainfall-runoff using conservation of mass and momentum distributed across multiple watersheds. This approach relies on characterizing the physics of infiltration, soil moisture, rainfall, and runoff rather than on a human forecaster inserted into the river forecast loop. In Case Study I— Reservoir Inflow Forecasting, the operational simulation of local inflow is a key component necessary for efficient power generation and flood management at upstream locations along the impoundment and in downstream river reaches. Upstream inflow from a river controlled by a flood control reservoir and contributions from an uncontrolled watershed are combined for making gate operation decisions for a dam in central Missouri. A hydrometeorological network is used to measure current precipitation, which is combined with numerical weather forecasts for hydrologic predictions out to 7-days. Continuous calibration of the model using continuous simulation trials, rather than event simulation, helps refine model parameters and to understand sensitivity of soil moisture to PET estimation methods. Through this procedure two infiltration parameters controlling runoff were fine tuned for a recent period containing a range of event magnitudes, from low to high discharge that is critically important for the required application.

Distributed flood forecasting in an urban drainage context is demonstrated with several events constructed from radar-based rainfall over a 260 km^2 watershed in Houston, Texas. In *Case Study II—Urban Flood Forecasting*, the influence of radar rainfall input uncertainty is illustrated. Removing the systematic error (bias) in GARR is essential to a real-time flood forecasting system. Bias-corrected rainfall-runoff is presented for a recent event showing close agreement, especially in the rising limb (also illustrated in Chap. 8 for Brays Bayou historical flood events). With accurate rainfall input, the full efficiency of the distributed model is realized as demonstrated by operational results. Distributed model soil infiltration parameters incorporate imperviousness and channel hydraulics composed of concrete-lined trapezoidal cross-sections are important model characteristics.

Assessing likely impacts of climate change is investigated with a physics-based model calibrated to a historical period. In Case Study III, climate change projections are made by applying perturbations derived from a GCM. The ratios of PET and precipitation for two periods, the last 30-year period in the twentieth and twenty-first century, are applied as a *delta* change made to a historical period of radar-based precipitation and PET, *as-if* the climate has already changed. Then under perturbed (future) climate, distributed streamflow is taken at locations of ecological interest, i.e., where fish species sampling exists in the Canadian River watershed spanning relatively arid to more humid climates across parts of New

Mexico–Texas–Oklahoma. A diminished summer streamflow climatology is projected in July, which may have important impacts on certain fish species present in the river. These case studies illustrate the use of a physics-based distributed approach for a range of hydrologic analyzes.

References

Bedient, P.B., A. Holder, J.A. Benavides, and B.E. Vieux. 2003. Radar-based flood warning system—tropical storm Allison. Journal of Hydraulic Engineering 89(6): 308–318.

Fang, Z., P.B. Bedient, and B. Buzcu-Guven. 2011. Long-term performance of a flood alert system and upgrade to FAS3: A Houston, Texas, case study. *Journal of Hydrologic Engineering* 16 (10): 818–828.

Farnsworth, R.K., E.S. Thompson, and E.L. Peck. 1982. *Evaporation atlas for the contiguous 48 United States*. National Oceanic and Atmospheric Administration, National Weather Service: US Department of Commerce.

Hamlet, A.F., and D.P. Lettenmaier. 1999. Effects of climate change on hydrology and water resources in the Columbia River Basin. *Journal of the American Water Resources Association* 35(6): 1597–1623.

Institute of Numerical Mathematic (INM). 2008. World Climate Research Programme (WCRP) Coupled Model Intercomparison Project phase 3 (CMIP3) Project Database: INMCM3.0 at Institute of Numerical Mathematics, Russian Academy of Science, Russia. NCAS British Atmospheric Data Centre. http://catalogue.ceda.ac.uk/uuid/67074e181159ad29ead1ccf07 e04517d. Accessed 9 March 2016.

Intergovernmental Panel on Climate Change (IPCC). 2007. Climate Change 2007: The Physical Science Basis. Contribution of Working Group I to the Fourth Assessment Report of the Intergovernmental Panel on Climate Change. Solomon, S., D. Qin, M. Manning, Z. Chen, M. Marquis, K.B. Averyt, M. Tignor, and H.L. Miller (eds.). Cambridge University Press, Cambridge, United Kingdom and New York, NY, USA, 996 pp. http://www.ipcc.ch/ ipccreports/.

Juan, A., Z. Fang, and P.B. Bedient. 2015. Developing a radar-based flood alert system for sugar land, Texas. *Journal of Hydrologic Engineering* E5015001.

Looper, J.P., and B.E. Vieux. 2012. An assessment of distributed flash flood forecasting accuracy using radar and rain gauge input for a physics-based distributed hydrologic model. *Journal of Hydrology* 412: 114–132.

Looper, J.P., Vieux, B.E. and Moreno, M.A., 2012. Assessing the impacts of precipitation bias on distributed hydrologic model calibration and prediction accuracy. *Journal of Hydrology*, 418: 110–122.

Maurer, E.P. 2007. Uncertainty in hydrologic impacts of climate change in the Sierra Nevada, California under two emissions scenarios. *Climatic Change* 82: 309–325.

Meehl, G.A.C., T. Covey, M. Delworth, B. Latif, J.F.B. Mcavaney, R.J.Stouffer Mitchell, and K.E. Taylor. 2007. The WCRP CMIP3 multimodel dataset—a new era in climate change research. *Bulletin of the American Meteorological Society* 88(9): 1383–1394.

Miller, N.L., K. Bashford, and E. Strem. 2003. Potential climate change impacts on California hydrology. *Journal of the American Water Resources Association* 39(4): 771–784.

Oudin, L., C. Perrin, T. Mathevet, V. Andreassian, and C. Michel. 2006. Impact of biased and randomly corrupted inputs on the efficiency and the parameters of watershed models. *Journal of Hydrology*. doi:10.1016/j.jhydrol.2005.07.016.

Reclamation. 2008. Sensitivity of Future CVP/SWP Operations to Potential Climate Change and Associated Sea Level Rise, Appendix R in Biological Assessment on the Continued Long-term

Operations of the Central Valley Project and the State Water Project, prepared by Bureau of Reclamation, 134. U.S. Department of the Interior, August 2008.

Reclamation. 2009. Sensitivity of Future Central Valley Project and State Water Project Operations to Potential Climate Change and Associated Sea Level Rise, attachment to Appendix I in Second Administrative Draft Program EIS/EIR—San Joaquin River Restoration Program, 110.

Reclamation. 2010. Climate Change and Hydrology Scenarios for Oklahoma Yield Studies. Technical Memorandum 86-68210-2010-01, Technical Service Center, Water Resources Planning and Operations Support Group, Water and Environmental Resources Division. April 2010, 59.

Rosenfeld, D., D.B. Wolff, and D. Atlas. 1993. General probability-matched relations between radar reflectivity and rain rate. *Journal of Applied Meteorology* 32: 50–72.

Smith, M.B., Koren, V., Zhang, Z., Zhang, Y., Reed, S.M., Cui, Z., Moreda, F., Cosgrove, B.A., Mizukami, N. and E.A. Anderson. 2012. Results of the DMIP 2 Oklahoma experiments. *Journal of Hydrology*, 418: 17–48.

Vieux, B.E., and P.B. Bedient. 2004. *Assessing urban hydrologic prediction accuracy through event reconstruction.* Special Issue on Urban Hydrology: Journal of Hydrology.

Vieux, B.E., C. Chen, J.E. Vieux, and K.W. Howard. 2003. Operational deployment of a physics-based distributed rainfall-runoff model for flood forecasting in Taiwan. In *Proceedings*, eds. Tachikawa, B. Vieux, K.P. Georgakakos, and E. Nakakita, International Symposium on Weather Radar Information and Distributed Hydrological Modelling, IAHS General Assembly at Sapporo, Japan, July 3–11, IAHS Red Book Publication No. 282: 251–257.

Vieux, B.E. and Moreda, F.G., 2003. Ordered Physics-Based Parameter Adjustment of a Distributed Model. *Calibration of Watershed Models*, pp.267–281.

Vieux, B.E., 2004. Distributed hydrologic modeling. *Distributed Hydrologic Modeling Using GIS*, Springer: pp.217–238.

Vieux, B.E., P.M. Thompson, and J.E. Vieux. 2015. Forecasting the Magic Dragon: How Much Inflow Will There Be, and When Will It Arrive in Lake of the Ozarks? Conference Proceedings, HydroVision International, July 14–17, Portland Oregon.

Walter, I.A., R.G. Allen, R. Elliott, M.E. Jensen, D. Itenfisu, B. Mecham, T.A. Howell, R. Snyder, P. Brown, S. Echings, and T. Spofford. 2000. ASCE's standardized reference evapotranspiration equation. In Proceedings of the Watershed Management 2000 Conference, June.

Young, C.B., A.A. Bradley, W.F. Krajewski, A. Kruger, and M.L. Morrissey. 2000. Evaluating NEXRAD multisensor precipitation estimates for operational hydrologic forecasting. *Journal of Hydrometeorology* 1(3): 241–254.

Chapter 12
V*flo*®—Software for Distributed Hydrology

Abstract V*flo*® is a distributed model for hydrologic prediction and analysis that is based on: recently developed distributed modeling techniques; multisensor precipitation estimation; and secure client/server architecture. It utilizes GIS and remotely sensed data. Developed from the outset to utilize multisensor inputs, this model is suited for distributed hydrologic prediction in post-analysis and online for continuous forecasting. The hallmark of V*flo*® is its prediction of flow rates and stage in every grid cell defined by the hydraulics of overland and channel flow. An integrated network-based hydraulic approach to watershed modeling has advantages that make it possible to represent both local and main-stem flows with the same model setup and simultaneously. This integrated approach is useful for urban and natural watershed applications, reservoir inflow forecasting, flood prediction, and hydrologic analysis of land use and climate change.

12.1 Introduction

The V*flo*® software is presented, as an online operational version, and a desktop version with a graphical user interface that supports setup, sensitivity study, animation, statistical network analysis, and model calibration. Various precipitation input options include rain gauges, and radar formats typical of the WSR-88D including NEXRAD Level II and Level III products. Cells comprising the drainage network can be overland, trapezoidal channel, defined by cross-sections input from surveys or extracted from a high-resolution DEM, base cells, or reservoir cells with gated outflow. Inundation predictions rely on stage in channels distributed within the basin, which is then distributed based on gridded elevation differences within the floodplain adjacent to the channel. Output from controlled reservoirs is modeled by including discharge-operating rules. Uncontrolled reservoirs are modeled by incorporating stage-discharge and stage-volume relationships.

In real time, the model is a 'hot-start' model such that the simulation uses the last solution and continues without having to go back to a starting time. In general, real-time operations are faster than post-analysis. Computation time may vary

© Springer Science+Business Media Dordrecht 2016
B.E. Vieux, *Distributed Hydrologic Modeling Using GIS*,
Water Science and Technology Library 74, DOI 10.1007/978-94-024-0930-7_12

depending on how many monitoring points are saved to the database and other configurable options. As the number of cells increases, the increase in CPU time is much less than 1:1. The progressive efficiency of the model supports scales from catchment or river basins, to entire regions or countries. Current installations of V*flo*® demonstrate that a fully distributed physics-based model is capable of real-time operation for basins as large as 32,000 km². Larger domains are possible depending on memory and number of watch points set for hydrograph generation. Hydrographs can be produced at any junction in the network formed by the finite elements connecting each grid cell, called a watch point. The connectivity of the drainage network and its hydraulic characteristics define the hydrograph shape. The user may additionally select output of gridded stage for analyzing inundation at each time interval solved.

Within the V*flo*® GUI, query, display, and control of drainage network connectivity, parameters, and rainfall inputs are supported for analysis purposes and preparation of the model for real-time flood forecasting. Figure 12.1 shows a 100-m resolution drainage network delineated from a 3-m resolution DEM. The network consists of overland and channel cells that flow into a reservoir (lake shown in dark blue with red dot indicating a stream gauge at the reservoir outlet). The rainfall total shown is for an extreme event that occurred over the Colorado Front Range in Sept 8–16, 2013. This extreme event totaled 210 mm (8.3 in.) in 192 h over this watershed.

Fig. 12.1 Drainage network produced by V*flo*® at 100-m resolution with storm total rainfall exceeding 210 mm (8.3 in.) over a period of 192 h

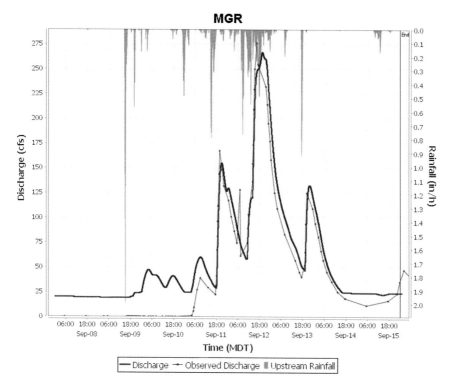

Fig. 12.2 Simulation of reservoir level and resulting gated outflow from the reservoir ($1 \text{ m}^3\text{s}^{-1} = 35.31$ cfs)

Simulation of reservoir level and interior stream gauge locations in the Desktop version support model calibration and setup for real-time operations. In Fig. 12.2, the runoff from this event is routed through the reservoir with hydraulic rating curves representing gate operations. After model calibration to this event at an upstream gauge location, simulated and observed discharge hydrographs are seen to match closely.

The physics-based approach to hydrologic modeling that forms the basis for V*flo*® is presented in Vieux (1988), Vieux et al. (1990), Vieux (1991), Vieux and Gauer (1994), Vieux and Vieux (2002); and Vieux et al. (2008). Applications of V*flo*® in stormwater, flood forecasting, and watershed management are described by Bedient et al. (2003, 2013), Doubleday et al. (2013), Fang et al. (2006, 2009), Hunter et al. (2003), Kim et al. (2008), Looper et al. (2012), Looper and Vieux (2012), Noh et al. (2012), Oh et al. (2010), Robinson et al. (2009), Rojas-González et al. (2008, 2010), Safiolea et al. (2005), Teague et al. (2013), Torres Molina et al. (2013), Vieux and Vieux (2002), Vieux et al. (2003, 2005, 2009); and for recharge and water balance in Moreno and Vieux (2013). Software development has added capabilities helping to accomplish a variety of distributed hydrologic modeling objectives described in the following sections.

12.2 Building Watershed Models

In V*flo*®, the basin overland properties (BOP) file is the basin model. Besides storm event precipitation or observed flow, all elements of the watershed model are stored as a BOP file (**.bopx*). The BOP file stores the essential data that makes up a V*flo*® model: information on the model domain, drainage network, channel and reservoir locations, and model parameters such as roughness and hydraulic conductivity. BOP files also store information entered in the V*flo*® model after initial model development, such as: metadata, the location of watch points, calibration factors, and additional data such as shapefiles for map display. There are four key options for developing a basin model:

1. The "AutoBOP" software may be used to develop a drainage network, establish the V*flo*® domain, and import parameter maps. A high-resolution DEM is required. Channel and basin boundary shapefiles are recommended for enforcing channels and watershed boundary. Maps of parameters such as roughness or infiltration may be developed externally in GIS, or set as spatially constant values. Geographic projection support is available to assist with re-projecting maps in disparate projections into a single target projection within AutoBOP.
2. A basin model may be manually developed within V*flo*® by drawing flow directions connecting upstream grid cells to downstream cells. Parameters can then be loaded as geographic map layers of hydraulics and infiltration panels are used to designate model parameters such as roughness and hydraulic conductivity.
3. All cell properties, as grids representing D8 flow directions and parameters, can be produced externally with GIS software and imported to create a BOP file in V*flo*®. A high-resolution digital elevation model (DEM), gridded soil maps, gridded land use maps, and other GIS data are required for this method.

Model creation using AutoBOP supports creation of any model resolution and percentage of channel cells thus creating a range of resolutions and channel network definition. Figure 12.3 shows the outlet selection step in AutoBOP, displaying delineated channel cells at the target model resolution. After selection of the outlet, model parameters are assigned, along with geomorphic channel-width, and channel cross-sections extracted from the DEM. The final step of AutoBOP loads these properties into V*flo*® for further configuration, calibration, or analysis.

12.3 Precipitation Input

12.3.1 Design Storm Analysis

Design storms can be modeled as static or dynamic (moving) hyetographs using the Storm Builder Extension. The design storm temporal distribution can be entered in,

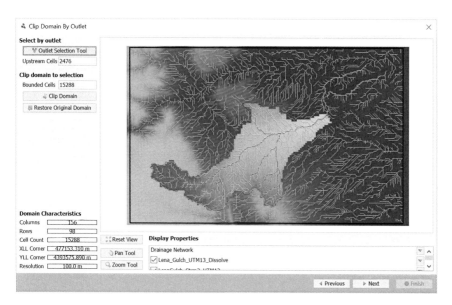

Fig. 12.3 AutoBOP model creation showing outlet selection and watershed boundary and stream channel delineation

or built-in SCS I, II, or III hyetographs can be specified. Applying a spatially uniform hyetograph will produce the same rainfall intensity at the same time over the domain. Storm Builder represents a moving storm as a storm that passes over the domain at specified speed and direction. Each grid will receive the same depth, but at different times related to the motion of the storm front. Vieux and Vieux (2010) present the methodology for measurement of storm characteristics from radar and rain gauge. Development of representative hyetographs from rain gauge and depth area reduction factors from radar is demonstrated in Vieux et al. (2015, 2016). Effect of storm direction and speed on flood response is shown by Carter and Vieux (2012). Figure 12.4 shows the Storm Builder Interface with the storm motion vector (to the northeast) and rainfall total indicated.

The hyetograph may be specified based on synthetic design storm parameters or developed from locally measured storm hyetographs. The hyetograph shown in Fig. 12.5 represents a front-end loaded storm typical of the Colorado Front Range, called the CUHP hyetograph (USDCM 2016). When the user inputs the temporal distribution of rainfall depths (hyetograph totals to 1.0), as seen in Fig. 12.6, it is applied with a specified depth, e.g., a 100-year return frequency. The hyetograph and total depth are then moved over the targeted watershed at a selected speed and direction. These moving intensities are applied to each model grid according to the specified motion vector, seen in Fig. 12.4. The motion vector can be applied to the target watershed with any given speed and direction. Preferably, the speed and direction of the local storms climatology should be used. However, more critical

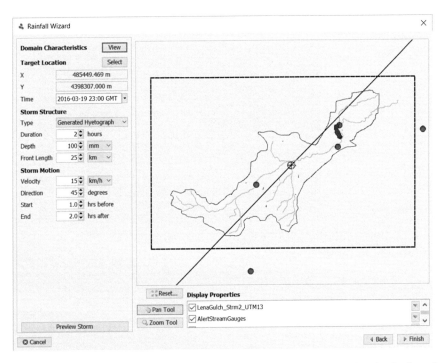

Fig. 12.4 Storm Builder interface with storm motion vector, speed, front length, depth, and duration (1 in. = 2.54 cm) (Vieux and Vieux 2010)

speeds and directions can be investigated with Storm Builder, for example, a slow moving storm in the same direction as the major drainage.

The effects of moving the hyetographs upslope or downslope in the watershed is demonstrated by applying the dynamic design storm in two directions, 45° (NE) and 225° (SW) corresponding to the general drainage direction of the target basin. The hyetographs sampled at four locations from upstream to downstream in the watershed result in the progressive delay of the same hyetograph (note there are some sampling differences, but they all total to 99 mm in 2 h (3.9 in.). From top to bottom, the peak arrival time is 12:45, 12:55, 13:10, and 13:25 seen in Fig. 12.7. Along the motion vector, the 10 and 15 min delays are produced by the storm moving at 15 km/h at four locations that are separated by 2.5 km and by 3.75 km apart. By changing direction of motion in Storm Builder, the hydrologic response can be examined. Peak discharge is magnified by the downslope-moving storm direction (45°), whereas the response is diminished by the upslope moving storm (225°), as shown by Fig. 12.8. It is interesting to note that the static storm hydrograph, produced by the same hyetograph but without movement, produces a peak of about 258 m^3s^{-1} (not shown), which is intermediate between 297 and 215 m^3s^{-1} peak discharge generated by downslope and upslope moving storms, respectively. The static storm with uniform spatial distribution is *not* the most

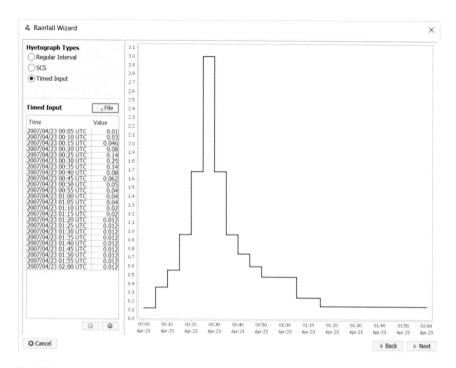

Fig. 12.5 Instantaneous rainfall intensities temporal distribution defined by the user in Storm Builder

conservative from a design view point since its peak discharge magnitude is less than the downslope-moving storm.

12.3.2   *Continuous Simulation*

The extension called *Continuous Simulator* provides a method for modeling long-term runoff with continuous input. Continuous Simulator supports solution of rainfall input for long time periods, extending from days to years. Soil moisture accounting is tracked continuously in a single-layer soil subject to evapotranspiration, infiltration, and rainfall with wetting front redistribution. Potential evapotranspiration is input then modified according to available rainfall and soil moisture in each model grid. Output includes instantaneous flow rate, average flow rate over time step, average depth of flow over drainage area upstream of watch point, and total cumulative depth upstream.

Event-based simulations are not greatly affected by ET during the event. During continuous simulations, however, soil moisture is depleted by actual ET and can have a significant effect on runoff volume and initiation of hydrographs. For the

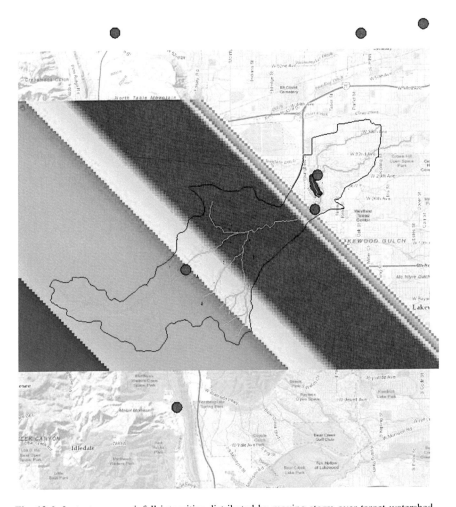

Fig. 12.6 Instantaneous rainfall intensities distributed by moving storm over target watershed

longer durations this can be modeled with the Continuous Simulator extension. Time-variable evapotranspiration (ET) may be entered as either climatological values for an annual cycle, or as a time series of values that are representative for the period simulated. If climatological values or a time series are to be used, be sure to include sufficient ET values to cover the period of simulation in order to avoid unexpected results. If only one year is included, then the model will repeat those values yearly for the entire simulation period. Continuous simulation results can be seen in the viewer shown in Fig. 12.9, for a three-month period. Soil moisture is also stored for retrieval, which may be used in refining event-based calibration results and understanding whether saturation affects runoff during specific event periods. Note the major event September 8–16, 2013, after a relatively dry period since July 1, 2013.

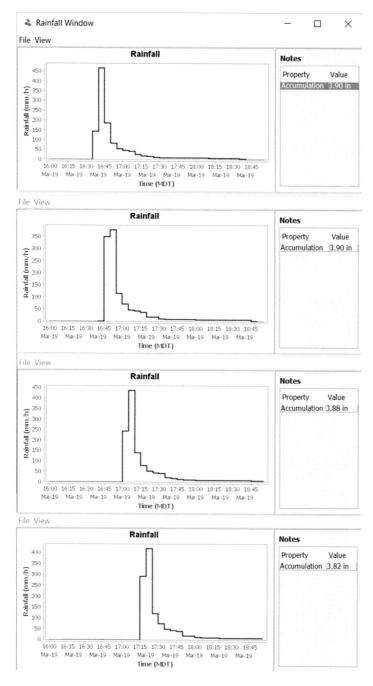

Fig. 12.7 Moving hyetographs for the downslope-moving storm (45°) sampled

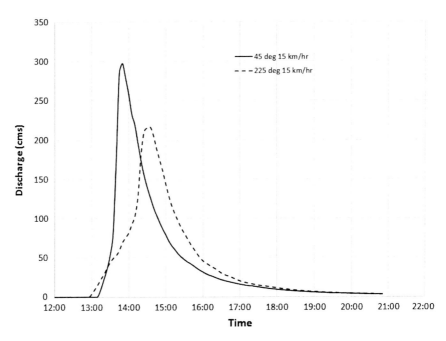

Fig. 12.8 Hydrograph response obtained by both downslope and upslope moving storms

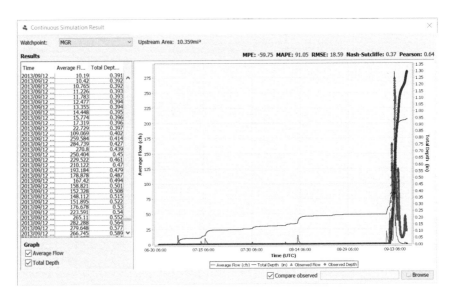

Fig. 12.9 Continuous result viewer showing discharge and cumulative runoff depth

12.4 Pipes Extension

Pipes Extension is provided in V*flo*® Desktop for modeling the influence of storm sewers as a secondary drainage network diverting flow from the primary overland and channel network. For example, diversions to detention basins or tunnels can be modeled using hydraulic inlet rating curves, percent diversion, or fixed discharge. Geometry of the pipe network and inlet locations can be imported as a simple text file containing coordinates and pipe length and diameter, or directly from the Storm Water Management Model (SWMM) pipe-node geometric configuration. The drainage network in V*flo*® is shown in Fig. 12.10, where the SWMM pipe network has been imported to represent the storm sewer network. A curb inlet was inserted into a selected model grid, and based on input parameters, flow is taken from the primary drainage network based on limiting components, either the inlet, or downstream pipe capacity. Surcharging of pipes is not modeled, so the limitation of a particular pipe is its full flow capacity.

12.5 Inundation Analyst

Inundation Analyst is a V*flo*® extension that provides images, animations, and data showing the extent of forecast or simulated inundation, which is an indication of flood risk. The extension's inundation products are especially useful for flood management applications: forecast inundation is useful for operational decisions, warning and notification, and coordinating emergency response. Inundation Analyst

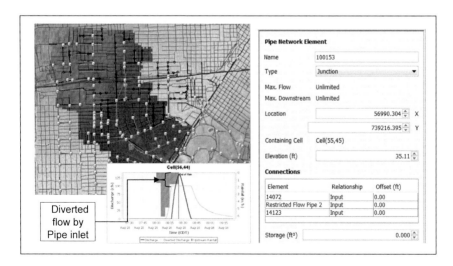

Fig. 12.10 Modeling the effect of hydraulic inlet diversion on surface runoff

operates independently from the V*flo*® model, but can use data exported from V*flo*® as input for generating inundation forecasts.

All input data must be in ESRI ASCII or binary grid format. Inundation Analyst requires a digital elevation model (DEM), a flow direction map, a channel flow direction map, and stage data. If flow direction maps are in a different resolution than stage data, they will be filtered automatically. Inundation Analyst provides functionally used to analyze a DEM to delineate the flow direction used with Inundation Analyst to map inundated areas and depths. Stage data inputs are those exported from a V*flo*® model. The grid definition of the flow direction map extracted from a high-resolution is necessary for mapping stage files produced at the V*flo*® model grid resolution, which is usually coarser. For example, a LiDAR DEM at 5-m resolution could be used to prepare a flow direction grid for use in Inundation Analyst by mapping exported stage grids from V*flo*® produced at 100-m resolution in the same or different different geographic projection.

A high-resolution flow direction map created for use in Inundation Analyst from DEM files is presented in Fig. 12.11. The high-resolution flow direction grid (in this case, at 10-m) is used to interpolate inundation depth produced at coarser resolution (shown here at 100 m) by V*flo*®. Once created, the Inundation Analyst project can be saved to disk with the extension, ".*indx*," for later use. Once all of the images have been produced, the first image will be shown in the Inundation Analyst GUI.

The *Animation* menu contains the tools necessary to generate and control animation sequences. Each map of stage exported from V*flo*® is displayed as an

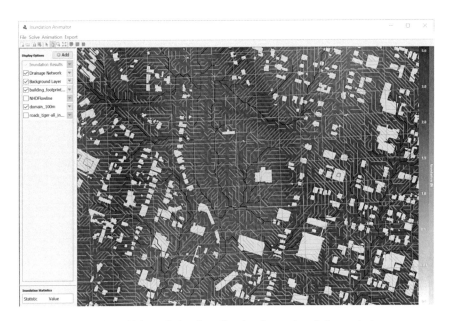

Fig. 12.11 Processing a high-resolution flow direction for use inundation analyst

Fig. 12.12 Inundated depth (*blue*) mapped with buildings whose elevation is below the water surface elevation (*red*) at a particular point in time during a flood

overlay with other shapefiles, for example building locations. Evaluation of which building is flooded or not flooded can be produced, as shown in Fig. 12.12. Maximum inundation depth or at selected times may be evaluated for display of inundated buildings or other structures relative to the water surface elevation, and related statistics. Export of the inundation maps is possible as images, ESRI ASCII grids, or map formats such as shapefiles or KMZ.

12.6 Model Feature Summary

The V*flo*® GUI provides dropdown menus with analysis options and tabs showing modules for controlling model operation or setup. Two main components are stored separately as: (1) basin overland properties (BOP), and (2) rainfall (RRP) files. The BOP file contains information related to terrestrial characteristics such as drainage direction, slope, hydraulic roughness, infiltration, and calibration factors. The RRP file contains the event-specific rainfall rate information and is optimized for loading and computation. RRP files may be generated from a variety of sources including rain gauge, radar, or other multisensor precipitation estimates. Modules are displayed according to the mode selected, or depending on what information has been loaded, e.g., precipitation will only be displayed once RRP files for an event or a design storm have been loaded. A summary is presented in Table 12.1 of the features provided by V*flo*® version 6.1.

Table 12.1 V*flo*® features summary

Feature description and type	Feature type
GIS shapefile and background image loading and display	G
Rainfall input Display: generate cell hyetographs or for selected cells/subbasins	G
Rainfall Input Display: animation of rainfall at each time interval	G
Display parameter values, georeferenced background image, rainfall total, and image export	G
Time zone, geographic projection, and metadata assignment	G
Pan, zoom, draw flow direction, and selection of cell types or stream reach	G
AutoBOP Wizard—DEM processing for watershed delineation, slope assignment, % stream channel selection, and parameter assignment	P
Baseflow assignment by stream reach	P
Parameter Input Wizard support for import gridded maps, or assigning constant parameters	P
Green-Ampt infiltration parameters including soil depth	P
Calibration—slider bars applied to parameters in selected grid cells	P
Import/Export of gridded parameters	P
Save/load rating curves and cross-sections for entire domain	P
Save/load observed hydrographs for entire domain	P
Export domain shapefile of model grid and of watch point locations	P, G
Gridded flow or stage export for entire domain	A
File-based Save/Load Rating Curve and Cross-sections for entire domain	A
View Result Table of solved hydrographs, performance statistics, peak, and volume	A, G
Network Statistics—area, cell numbers, and rainfall and parameter histograms	A, G
Analyze and display celerity by reach for identification of problem areas	A, G
Import/Export of hydrograph stage and discharge and hyetograph text files	A
Import/Export stage-storage and stage-discharge reservoir curves	A
Import/Export stage-area and discharge-stage stream gauge rating curves	A
Watch points assignment for solving and hydrograph display	A
Assigning observed hydrographs as inflow boundary conditions	A
Cell types: base, overland, channel, rated channel, cross-section, reservoir detention, and body	A
View Rated Channel Cells and Cross-section Cells	A, G
Kinematic Routing fourth-order accurate solver	A
Pipes module for defining hydraulic inlet diversion to a pipe network	A
Storm Builder—dynamic storm generation tool	A
Sensitivity Module (Sensi)	A
AutoCal Wizard for exploring parameter sensitivity for multiple events	A, G
Evapotranspiration assignment with gridded/constant crop coefficients	A
Inundation Analyst maps stage and evaluates flooded structures	A
Precipitation filtering gridded/polar products into model grid (BAG Maker)	A

(continued)

Table 12.1 (continued)

Feature description and type	Feature type
Continuous Solver support for solving long-term rainfall input	A
Continuous Solver project results for display of soil moisture and runoff hydrographs	A
Rainfall Wizard for loading gridded rainfall maps, producing uniform hyetographs, interpolation from rain gauge networks, and loading previously saved RRP input	A

Feature Type Graphical (*G*), Parameter (*P*), Analytical (*A*)

12.6.1   Cell Types

V*flo*® contains seven different cell types: overland, channel, rated channel, cross-section channel, reservoir, reservoir body, and base cells. Each of these cell types has characteristics which separate them from other cell types. Runoff generated by V*flo*® is determined by cell type and cell properties. Each cell type is used to model explicitly the drainage network components controlling runoff. The Cell Types available are:

Base Cell
: Water is propagated according to the kinematic wave speed $c^2 = gh$, where g is the acceleration due to gravity and h is the water depth and is time variable depending on discharge rate in the cell and has no parameters. It is the default cell, but may be used to model shallow water, wetlands, and upper reaches of reservoirs.

Reservoir Body Cell
: Water is also propagated according to the kinematic wave speed c, but the depth depends on the difference between a grid cell bottom elevation and downstream lake level. An outflow boundary condition must be specified in the downstream cell representing gate operations or set outflow rate.

Overland Flow Cell
: Normal depth is governed by Manning's equation assuming uniform flow depth over the grid cell.

Channel Cell
: Conveyance is computed as the sum of the trapezoidal cross-section and overland flow depth within the cell. The channel-width is not associated with the cell width. Trapezoidal hydraulic parameters include base width, side-slope, roughness, and bedslope. Trapezoidal cross-sections may be applied in between cross-section or rated channel cells.

Rated Cell   Flow is determined by complex hydraulics due to compound
                          cross-section, roughness elements, or hydraulic structures.
                          Discrete pairs of stage-area and stage-discharge are entered
                          to characterize the hydraulic performance of a cross-section.
Cross-Section Cell   Measured cross-section data describing channel geometry
                          (station and distance) may be input along with hydraulic
                          roughness and slope to synthesize the stage-discharge
                          relationship.
Reservoir Cell   Uncontrolled reservoirs affect outflow at a cell through a
                          stage-volume and stage-discharge. For reservoirs longer
                          than one cell, reservoir body cells may be used to propagate
                          the flood wave from an upstream channel cell to the cell
                          containing the reservoir. Controlled reservoirs are input as
                          boundary conditions at the outlet cell.

12.6.2 Network Statistics

V*flo*® provides the ability to analyze the upstream network statistics of specific
cells. Selecting a cell, and pressing F1 (or activating the *Analysis | Network
Analysis | Analyze Cell* menu item) can generate a statistical report on the network
connected to the selected cell. A window will appear entitled Network Statistics.
Items listed under *Cell Information* include cell name, grid coordinate, number of
connected cells, drainage area, and the average number of cells that flow into a
downstream cell related to the complexity of the network. The Drainage Area is the
number of connected cells multiplied by the area of each grid cell. Besides
parameter statistics, precipitation depth over the area that drains to the cell is
computed for this display. Comparing this depth with the streamflow volume is
useful for calibration studies and documenting input and output at a specific cell
location. The resulting information, shown in Fig. 12.13, is useful for summarizing

Network Analysis   —   □   ×

Network Properties   Cell Parameters   Parameter Histogram

Property	Value		Parameter	Min.	Avg.	Max.
Selected Cell	Cell(75,279)		Celerity (ft/s) (stage=5.0 ft)	0.05	75.31	257.87
Upstream Cells	12475		Roughness (Ratio)	0.0100	0.0197	0.2000
Connectivity	1.9		Slope (%)	0.001	3.300	23.833
Drainage Area	17.34 mi²		Hydraulic Conductivity (in/hr)	0.00	0.07	0.96
Total Baseflow	0.238 cfs		Wetting Front (in)	0.00	26.94	48.31
Number of Base Cells	0		Porosity (Ratio)	0.00	0.19	0.38
Number of Overlan...	11715		Soil Depth (in)	0.0	41.8	79.5
Number of Channel...	760		Initial Saturation (Ratio)	0.00	0.05	0.07
			Imperviousness (Ratio)	0.00	0.13	1.00
			Abstraction (in)	0.00	0.00	0.00
			Channel Width (ft)	0.57	35.10	123.00
			Channel Side Slope (Ratio)	1.00	1.00	1.00

Fig. 12.13 Statistics page shown for a selected drainage area

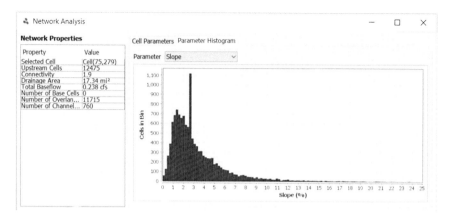

Fig. 12.14 Network statistics histogram of slope

average statistics and reporting and examining the combination of parameter value and calibration multiplier. Each parameter with spatial variation is plotted as a histogram seen in Fig. 12.14, shown here for slope.

For purposes of identifying excessively high slope or anomalous rating curves, a *Reach* tool is added to V*flo*® version 6.1.x that displays celerity for a selectable depth of flow. Abrupt changes in celerity are seen in Fig. 12.15, which can be caused by round-off or stage measurement intervals that differ for area and discharge. Figure 12.16 shows the Reach Analysis tool and associated channel cell on right with its rating curves. Examination of this plot allows the user to pinpoint locations that may be causing slow computations due to high-celerity and short time steps needed to model the hydraulics of flow. Difficulties with parameters and celerity can arise from:

- Rating curves that have abrupt increases in area accompanied by no or minimal change in discharge.
- Rating curves from modeled bridge hydraulics that have pressure flow when the water surface elevation rises above the low chord of the bridge or culvert.
- Channel grids produced from stream channel enforcement, causing a channel grid to be assigned a slope value that comes from the channel *sideslope*, usually much higher than bedslope value causing high-celerity.
- Errors in assigning model parameters or hydraulic properties even though checks are in place to limit values to typical ranges. A calibration multiplier can cause the parameter value that is within range to fall outside of physically realistic limits.

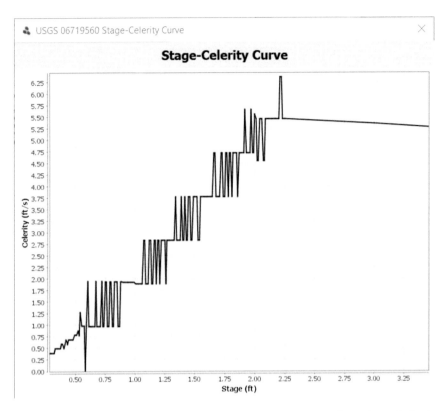

Fig. 12.15 Celerity curve showing oscillations due to round-off at different intervals in the stage-area and stage-discharge rating curves

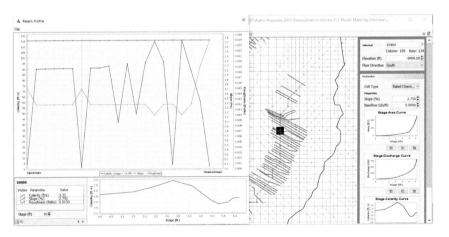

Fig. 12.16 Reach analysis tool for investigating abrupt changes in celerity and rating curve anomalies

12.6.3   Channel Routing

The governing routing equations used to solve the network may be toggled on/off by selecting under *Options | Analysis Options*. The following options are available for extending the kinematic wave model applicability to larger river systems using routing techniques that take into account the temporary channel storage and other terms in the full dynamic wave equations. Options for channel routing include:

- Modified Puls (storage indication)
- Observed flow (boundary condition)
- Looped rating curve modification
- Rating curves and cross-sections for complex geometry/hydraulics

The above example of rating curves generated from measured stream depth, area, and velocity demonstrates how sampling can result in unexpected behavior. Figure 12.17 shows field measurements for both area and discharge fitted to the data and then sampled at the *same* intervals of stage to avoid round-off, producing a smooth and monotonically increasing celerity curve, as opposed to Fig. 12.15 above.

12.6.4   Baseflow

Baseflow in units of m^3s^{-1} or cfs is applied by selecting an upstream channel cell, right clicking and selecting *Upstream Baseflow Start Point*. Next, select a downstream channel cell, right click and select *Downstream Baseflow End Point*. The user will be prompted to enter a total baseflow value (m^3s^{-1} or cfs) to be distributed as units of m^3s^{-1}/m between the upstream and downstream selected channel cells between the upstream and downstream selected channel cells.

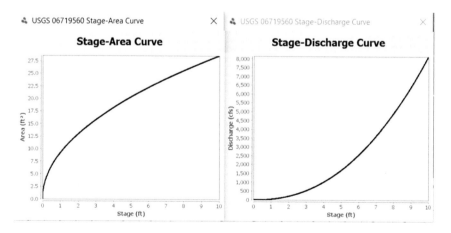

Fig. 12.17 Complex hydraulics represented as smoothly varying stage-area and stage-discharge

12.6.5   Infiltration

Two potential runoff processes are modeled in V*flo*®: (1) before saturation, when infiltration rate excess dominates; and (2) after saturation, when saturation excess dominates. Both may operate simultaneously in any given grid cell and spatially across the watershed. Once the soil profile becomes saturated, all rainfall runs off from that cell. The progression of the wetting front is modeled as piston flow, commonly known as the Green and Ampt infiltration routine. The wetting front suction, saturated hydraulic conductivity and degree of saturation are specified for a single-layer soil profile of variable depth. Incorporating the physics of this process makes the model sensitive to antecedent moisture conditions through degree of saturation.

In general, runoff generation is affected by rainfall rates exceeding infiltration rates. Once the available porosity in the soil profile is filled, then saturation excess runoff begins. The infiltration module accounts for the progression of a wetting front through a specified soil depth. This infiltration routine incorporates variable infiltration rates affected by antecedent conditions, soil properties, impervious area, initial abstractions, and soil depth. If Green and Ampt parameters are not available or only saturated hydraulic conductivity is known, a constant infiltration rate is applied.

Providing both saturation excess and infiltration excess offers the potential to model a wide variety of applications. Urban areas are assigned a percentage of the cell that is impervious. The impervious areas may be represented using a gridded parameter map containing cell values that range from 0 to 100 % imperviousness. The impervious fraction in a cell overrides the infiltration routine resulting in two conditions:

1. Rainfall rate less than infiltration rate: runoff is the product of rainfall rate and the impervious fraction.
2. Rainfall rate greater than infiltration rate: runoff is the difference between the rainfall and the infiltration rate not affected by the impervious area.

Green and Ampt parameters may be estimated from soil properties such as bulk density, and percentages of sand, silt, and clay, or based on soil texture. If only hydraulic conductivity is known or can be estimated, then V*flo*® treats the infiltration as a constant rate.

12.6.6   Precipitation Input Format

Precipitation input comes from spatially distributed rainfall rates derived from rain gauge, radar, or from already merged products. If a file format is known to V*flo*® (e.g., NEXRAD Level II or III) such maps are automatically filtered into the model grid when loaded. Alternatively, a filter file, a shapefile defining the spatial arrangement can be used to load an unknown format, provided ASCII files are provided with properly formatted time stamp as part of the file name. A Basin Average Group or BAG file can be created and used for customized import of rainfall.

12.6.7   Calibration

The model may be calibrated by loading precipitation maps for historical events, as along with observed stream flow data for these events. Calibration is achieved by comparing simulated versus observed volume/peak hydrographs. Infiltration, roughness, channel base width, channel side slope, baseflow, and rainfall are all parameters that may need to be adjusted by a calibration factor. These adjustments may be applied to an entire basin, subbasin, a few selected cells, or an individual cell. This is controlled by the available cell selection methods.

The calibration factor acts as a scalar multiplier to the base parameter value and is saved with the BOP file. The slider bars used to calibrate the watershed area selected as shown in Fig. 12.18. Through model calibration, simulated and observed runoff volume should first be matched, then timing and peak discharge addressed following the OPPA method described in previous chapters.

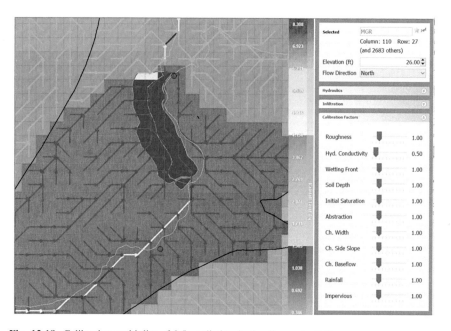

Fig. 12.18   Calibration multiplier of 0.5 applied to hydraulic conductivity

12.7   V*flo*® Real-time

The real-time edition of V*flo*® provides simulations and flood risk information in an unattended mode. The basin model created with the desktop is used in the real-time version by placing the calibrated BOP file on the server. Hydrographs at watch points set in the BOP file will be displayed on an associated webpage. Predicted stage is shown in Fig. 12.19 for an operational model that yielded a 3-h lead time, and achieved close agreement with observed stage within 0.15 m at its peak with over 12 m in depth and rising within 6 h. The resulting hydrograph shown is derived from gauge-adjusted radar rainfall (GARR), which uses rain gauges, plus QPF with an upstream boundary condition (stream gauge), shown in purple. Observed discharge is shown as gray-green dots.

Operationally, as rainfall files are input into the hydrological model, new hydrographs are calculated for each point. When new hydrographs are available, the webpage automatically updates the hydrographs and status icons. Below each hydrograph on the web page is information about the hydrograph, as well as links to see the forecast stage and discharge XML data from which the hydrographs are generated. This data can easily be retrieved for operational forecast applications and integration of model output for other applications.

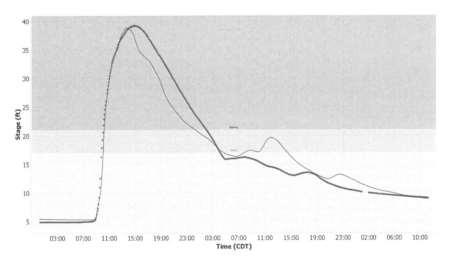

Fig. 12.19 Operational model that yielded more than 3-h lead time, and achieved close agreement with observed stage to within 0.15 m at its peak stage

12.8   Data Requirements

Data requirements are listed below for existing modules. As additional modules become available, the data requirements will be extended or revised. This outline provides a guideline for the basic information required to setup $Vflo^{®}$ for new basins.

1. Watch Points
   Selected prediction and for comparison to observed discharge. Locations of known stream gauges, reservoir inflow and outflow, control structures, overflow and diversions, and other critical locations. Discharge and stage elevations are generated at selected forecast points.
2. Reservoirs
   Uncontrolled reservoirs to be modeled require storage indication curves. Stage versus volume and discharge for known detention basins or reservoirs are needed to model storage effects. Reservoir body cells provide a method for routing through the water body whose outflow is controlled by gate operations.
3. River routing
   Kinematic wave analogy is used to route flow through channel cross-sections defined for each grid cell. Modified Plus may be applied to river reaches accounting for temporary storage volume between channel cross-sections throughout the channel network.
4. Channel cross-sections and hydraulic information
   Channel hydraulics are controlled by slope and roughness, and geometry defined by surveyed cross-sections or extracted from a high-resolution DEM. The trapezoidal cross-section provides high-celerity and should be used in conjunction with cross-sectional geometry especially where floodplains are present. Rating curves for measured or modeled can be set defining stage versus area and discharge. Geomorphic relationships may be used to relate channel-width to drainage area by inputting upstream and downstream width-drainage area.
5. Soils/Geological material maps for estimating infiltration
   Soil map, properties such as bulk density, porosity, texture classification, particle size distribution are used for deriving hydraulic conductivity and other Green and Ampt parameters. Maps of each parameter must be prepared separately for input.
6. Digital Terrain Model
   A high-resolution DEM is needed along with stream channels for enforcing delineation and watershed boundary for constraining drainage area. In well-defined topography that is relatively steep, enforcing streams and boundaries may not be necessary. Principal flow direction in some areas may require editing. Examples of areas that may not be well-defined by the DEM include braided streams, alluvial fans, flat areas, cutoff-meanders, multiple channels that may carry flow under certain conditions, or highly varied topography that is not properly resolved by the available DEM resolution.

7. Landuse/cover for estimating overland flow hydraulic parameter.
   Manning roughness can be developed from a GIS procedure that assigns a roughness coefficient to each landuse/cover classification in the map. Estimated or assumed values may be assigned to subbasins and then calibrated. The roughness map must be prepared separately for input.

In the absence of any particular dataset, except for the Digital Terrain Model, default parameter values may be used and then calibrated. Deriving parameter values from remotely sensed or geospatial digital data requires lookup tables that can be developed using the methods described in this book or from other expert sources.

12.9   Summary

V*flo*® is a fully distributed physics-based model for real-time and post-analysis prediction of rainfall-runoff. Using GIS data, the model is setup using the Desktop version for watersheds ranging from small upland catchments to larger river basins. With the model extensions and tools, it has the needed features for developing watershed models; calibration and validation; developing design storm analyzes; generating inundation maps; and investigating sources of runoff and water balance components over continuous periods of simulation. This model was developed from the outset to utilize multisensor inputs derived from radar. An integrated network-based hydraulic approach to hydrologic prediction has advantages that make it possible to represent both local and main-stem flows within the same model. The server edition with direct connection to radar and/or rain gauge inputs places V*flo*® model in a limited class of models that can perform operational prediction of distributed flow rates, hydraulic/hydrologic routing, and perform inundation mapping in real time.

Software
Additional materials are available on the software website including model description, tutorials and data sets. V*flo*® software may be downloaded with an evaluation license from: www.vieuxinc.com/getvflo

References

Bedient, P.B., A. Holder, J.A. Benavides, and B.E. Vieux. 2003. Radar-based flood warning system applied to tropical storm allison. *Journal of Hydrologic Engineering* 8(6): 308–318.
Bedient, P.B., W.C. Huber, B.E. Vieux. 2013. *Hydrology and floodplain analysis,* Fifth Edn., 801. Prentice-Hall, Inc., One Lake St., Upper Saddle River, NJ 07458. ISBN 0-13-256796-2.
Carter, P. and B.E. Vieux. 2012. Analysis of Storm movement and temporal distribution of rainfall and its influence on rainfall-runoff response of urban basins. 9th International Workshop on Precipitation in Urban Areas. St. Moritz Switzerland, 6–9 Dec 2012.

Doubleday, G., A. Sebastian, T. Luttenschlager, and P.B. Bedient. 2013. Modeling hydrologic benefits of low impact development: A distributed hydrologic model of the Woodlands, Texas. *JAWRA Journal of the American Water Resources Association* 49(6): 1444–1455.

Fang, Z., E. Safiolea, and P.B. Bedient. 2006. Enhanced flood alert and control systems for houston. In *Coastal Hydrology and Processes: Proceedings of the AIH 25th Anniversary Meeting & International Conference Challenges in Coastal Hydrology and Water Quality*, 199. Water Resources Publication.

Fang, Z., A. Zimmer, P.B. Bedient, H. Robinson, J. Christian, and B.E. Vieux. 2009. Using a distributed hydrologic model to evaluate the location of urban development and flood control storage. *Journal of Water Resources Planning and Management* 136(5): 597–601.

Hunter, S., B.E. Vieux, F. Ogden, J. Niedzialek, C. Downer, J. Addiego, and J. Daraio. (2003). A test of two distributed hydrologic models with WSR-88D radar precipitation data input in Arizona. 17th Conference on Hydrology, American Meteorology Society, JP3.3, Abstract and paper.

Kim, W.I., K.D. Oh, W.S. Ahn, and B.H. Jun. 2008. Study on flood prediction system based on radar rainfall data. *Journal of Korea Water Resources Association* 41(11): 1153–1162.

Looper, J.P., B.E. Vieux, and M.A. Moreno. 2012. Assessing the impacts of precipitation bias on distributed hydrologic model calibration and prediction accuracy. *Journal of Hydrology* 418–419: 110–122.

Looper, J.P., and B.E. Vieux. 2012. An assessment of distributed flash flood forecasting accuracy using radar and rain gauge input for a physics-based distributed hydrologic model. *Journal of Hydrology* 412: 114–132.

Moreno, M.A., and B.E. Vieux. 2013. Estimation of spatio-temporally variable groundwater recharge using a rainfall-runoff model. *Journal of Hydrologic Engineering* 18(2): 237–249.

Noh, H.S., B.S. Kim, N.R. Kang, and H.S. Kim. 2012. Flood runoff simulation using grid-based radar rainfall and V*flo*® Model. Paper HS276, in proceedings of the ERAD 2012, The Seventh European Conference on Radar in Meteorology and Hydrology.

Oh, K.D., S.C. Lee, W.S. Ahn, Y.H. Ryu, and J.H. Lee. 2010. A study on design flood analysis using moving storms. *Journal of Korea Water Resources Association* 43(2): 167–185.

Robinson, H., Z. Fang, and P. Bedient. 2009. distributed hydrologic model for flood prevention in the Yuna river watershed, dominican republic. In *Proceedings of World Environmental and Water Resources Congress 2009: Great Rivers*.

Rojas-González, A., E.W. Harmsen, and S. Cruz-Pol. 2008. Assessment predictability limits in small watersheds to enhance the flash flood prediction in western puerto rico. ASPRS 2008 Annual Conference Portland, Oregon, April 28–May 2.

Rojas-González, A.M.R., E.W. Harmsen, S.C. Pol, and Y.D.J. Arce. 2010. August. Evaluation of upscaling parameters and their influence on hydrologic predictability in upland tropical areas. In Proceedings of the AWRA Summer Specialty Conference, Tropical Hydrology and Sustainable Water Resources in a Changing Climate. San Juan, Puerto Rico.

Safiolea, E., P. Bedient, and B. Vieux. (2005) Assessment of the relative hydrologic effects of land use change and subsidence using distributed modeling. *Managing watersheds for human and natural impacts*, 1–7. ASCE. doi:10.1061/40763(178)87.

Teague, A., J. Christian, and P. Bedient. 2013. Radar rainfall application in distributed hydrologic modeling for cypress creek watershed, Texas. *Journal of Hydrologic Engineering* 18(2): 219–227.

Torres Molina, L.S., E.W. Harmsen, and S. Cruz-Pol. 2013, July. Flood alert system using rainfall forecast data in Western Puerto Rico. In Geoscience and Remote Sensing Symposium (IGARSS), 2013 IEEE International, 574–577. IEEE.

Urban Storm water design criteria manual (USWDCM). 2016. Urban Drainage and Flood Control District, Denver, Chapter 5—Rainfall. Vol 1.

Vieux, B.E.. 1988. Finite element analysis of hydrologic response areas using geographic information systems, 211 pp. Michigan State University.

Vieux, B.E. 1991. Geographic information systems and non-point source water quality modelling. Chichester, Sussex, England: John Wiley & Sons, Ltd. (*Journal of Hydrological Processes* 5: 110–123).

Vieux, B.E., V.F. Bralts, L.J. Segerlind, and R.B. Wallace. 1990. Finite element watershed modeling: one-dimensional elements. *ASCE, Journal of Water Resources Management Planning* 116(6): 803–819.

Vieux, B.E., and N. Gauer. 1994. Finite element modeling of storm water runoff using GRASS GIS. *Microcomputers in Civil Engineering* 9(4): 263–270.

Vieux, B.E. and J.E. Vieux. 2002. V*flo*®: A Real-time distributed hydrologic model. In *Proceedings of the 2nd Federal Interagency Hydrologic Modeling Conference*, July 28–August 1, 2002, Las Vegas, Nevada. Extended Abstract and CD-ROM.

Vieux, B.E., C. Chen, J.E. Vieux, and K.W. Howard. 2003. Operational deployment of a physics-based distributed rainfall-runoff model for flood forecasting in Taiwan. In *Proceedings, International Symposium on Weather Radar Information and Distributed Hydrological Modelling, IAHS General Assembly at Sapporo*. Ed. Tachikawa, B. Vieux, K. P. Georgakakos, and E. NakakitaJapan, July 3–11, 2003. IAHS Red Book Publication 282: 251–257.

Vieux, B.E., P.B. Bedient, and E. Mazroi. 2005. August. Real-time urban runoff simulation using radar rainfall and physics-based distributed modeling for site-specific forecasts. In 10th International Conference on Urban Drainage, 21–26. Copenhagen/Denmark.

Vieux, B.E., J.M. Imgarten, J.P. Looper, and P.B. Bedient. 2008. Radar-based flood forecasting: Quantifying hydrologic prediction uncertainty in urban-scale catchments for CASA radar deployment. Proceedings of *World environmental and water resources congress, Akupua'a*, 1–10. Honolulu, Hawaii: ASCE.

Vieux, B., J. Park, and B. Kang. 2009. Distributed hydrologic prediction: sensitivity to accuracy of initial soil moisture conditions and radar rainfall input. *Journal of Hydrologic Engineering*. doi:10.1061/(ASCE)HE.1943-5584.0000039,671-689.

Vieux, B.E. and J.E. Vieux (2010). Development of regional design storms for sewer system modeling. In *Proceedings of the Water Environment Federation, WEFTEC 2010: Session 87*, pp. 6248–6263(16).

Vieux, B.E., A. Dodd, and B. D. Wilson. 2015. Paper 11A.2 Radar analysis for design storm application. 37th Conference on Radar Meteorology, Sept 14–18, Norman, Oklahoma. Available on the Internet: https://ams.confex.com/ams/37RADAR/webprogram/Paper275331. html. Last Accessed, January 25, 2016.

Vieux, B.E., A. Dodd, and B. D. Wilson. 2016. Design storm temporal and spatial rainfall distributions from radar and rain gauge data analysis in nevada. Proceedings of ASCE/EWRI, West Palm Beach FL, May 22–26.

Index